Liposomes as Pharmaceutical and Nutraceutical Delivery Systems

Edited by Benjamin S. Weeks

Published in London, United Kingdom

Liposomes as Pharmaceutical and Nutraceutical Delivery Systems
http://dx.doi.org/10.5772/intechopen.1002158
Edited by Benjamin S. Weeks

Contributors
Abolfazl Mirani, Anielle Christine A. Silva, Arezoo Mohammadkhani, Auana R. da S. Andrade, Benjamin S. Weeks, Brett I. Pletschke, Carlo J.F. Oliveira, Christian I. Nkanga, Dhandara E. de L. Sampaio, Ehsan Kianfar, Fabiane C. de Abreu, Foued Salmen Espindola, Frank Ssengooba, Galilée G. Byamungu, Gauta Gold Matlou, Isabella de O.F. de Sousa, Jerusa M. Oliveira, Jonathan M. Mukanya, Juliana Reis Machado e Silva, Justin B. Safari, Jéssica M. Pereira, Larissa I.M. de Almeida, Lars Magnus Bergström, Livia M.S. de Lima, Lucas Anhezini, Marcos V. da Silva, Palesa Pamela Seele, Pathy B. Lokole, Paula Maseko, Pedro P. Perez, Rener M.F. Duarte, Rui W.M. Krause, Samaysa de L. Lins, Simnikiwe Nogqala, Tanaka Ndongwe, Virmondes Rodrigues, Xavier Siwe-Noundou, Zaid H. Mahmoud

First published in London, United Kingdom, 2025 by IntechOpen
IntechOpen is the global imprint of INTECHOPEN LIMITED, registered in England and Wales, registration number: 11086078, 167-169 Great Portland Street, London, W1W 5PF, United Kingdom

For EU product safety concerns: IN TECH d.o.o., Prolaz Marije Krucifikse Kozulić 3, 51000 Rijeka, Croatia, info@intechopen.com or visit our website at intechopen.com.

British Library Cataloguing-in-Publication Data
A catalogue record for this book is available from the British Library

Liposomes as Pharmaceutical and Nutraceutical Delivery Systems
Edited by Benjamin S. Weeks
p. cm.

This title is part of the Pharmaceutical Science Book Series, Volume 12
Topic: Pharmacology
Series Editor: Rosario Pignatello
Topic Editor: Cristina Manuela Drăgoi
Associate Topic Editor: Alina Crenguţa Nicolae

Print ISBN 978-0-85466-839-7
Online ISBN 978-0-85466-838-0
eBook (PDF) ISBN 978-0-85466-840-3
ISSN 3033-3318

IntechOpen Book Series

Pharmaceutical Science

Volume 12

Aims and Scope of the Series

Pharmaceutical science focuses on the design, synthesis, formulation, targeting, distribution, safety, and efficacy of active compounds as potential therapeutics. It is a large interdisciplinary discipline that aims to integrate the basic principles of physical and organic chemistry, biochemistry, biology, and engineering to discover, develop, and characterize active compounds and to optimize the formulation and delivery of drugs in the body for offering new and improved safe and efficacious therapies against human diseases. The research areas covered by the pharmaceutical sciences range from medicinal chemistry and pharmaceutical technology to pharmacology and toxicology, which represent the preliminary phases of drug development. Medicinal chemistry involves the design and synthesis of pharmaceuticals as well as the isolation of active agents from natural sources. Computer-aided strategies are increasingly involved in this drug discovery process. Pharmaceutics is a multidisciplinary science that examines the relationships between drug formulation, delivery, distribution, and clinical outcomes. Modern clinical approaches are increasingly relying on controlled release strategies and drug delivery and targeting systems, including nanotechnological platforms (nanomedicine). Pharmacology is the science of drug action in biological systems. Pharmacologists also make drugs as tools to explore aspects of cell and tissue functions. Toxicology is the study of the adverse effects of active agents on living organisms and the ecosystem, including the prevention and amelioration of such adverse effects. This book series includes volumes on Drug Discovery, Delivery, and Pharmacology. Their overall aim is to present the latest research in the whole path of drug discovery and development from different points of view of this multidisciplinary and dynamic field.

Meet the Series Editor

Prof. Rosario Pignatello is a Full Professor of Pharmaceutical Technology and Legislation at the University of Catania, Italy. He is the Director of the Department of Drug and Health Sciences. He has nearly 30 years of experience in the research and development of innovative formulations for the controlled release and targeting of bioactive molecules, through chemical approaches as well as nanotechnological carriers, aimed at treating different disorders. Prof. Pignatello has coauthored about 180 papers and edited a series of textbooks on biomaterials and their application in medicine. The main areas of his research are polymeric and lipid-based micro- and nanoparticles as modified drug delivery systems; vesicular nanocarriers (liposomes, micelles); lipophilic prodrugs and conjugates; synthesis and evaluation of new polymeric biomaterials for drug delivery and tissue regeneration. In particular, Prof. Pignatello works actively in the field of ocular drug delivery, leading the Research Centre for Ocular Nanotechnology, within the NANOMED Centre (Centre for Nanomedicine and Pharmaceutical Nanotechnology) at the University of Catania.

Meet the Volume Editor

Benjamin S. Weeks, Ph.D., earned his doctorate in molecular cell biology in 1988 from the University of Connecticut, Storrs, CT. His doctoral thesis focused on nutritional support of reproductive health in women. In 1988, Dr. Weeks joined the National Institute for Dental Research, Bethesda, MD, where he published studies documenting the mechanisms of nutrient and phytochemical support for the nervous and immune systems and wound healing. In 1995, Dr. Weeks joined the Division of Infectious Diseases at the University of Pennsylvania, Philadelphia, PA, as a Research Associate Professor of Medicine, where his work led to a breakthrough understanding of the anti-inflammatory effects of phytochemicals on viral infection. In 1997, Dr. Weeks joined the faculty of biology at Adelphi University, Garden City, NY and continued to publish over 80 peer-reviewed manuscripts focused on Vitamin C, policosanol, hemicelluloses, anxiolytic phytochemicals, selenium, and CBD. Dr Weeks' manuscripts have been published in journals such as *The Journal of Cell Biology, The Journal of Biological Chemistry, The Proceedings of the National Academy of Sciences, Fertility and Sterility, Obstetrics and Gynecology, Medical Science Monitor, Cellular Physiology, Immunology* and more. Dr. Weeks also has direct experience working with the nutraceutical industry. Between 2006 and 2009, Dr. Weeks demonstrated that nutrient extraction procedures patented by inventor Dr. Pedro Perez yielded nutrient products, PolicosnaolPlusTM, PureWayCTM and NatramuneTM, and CBD with enhanced bioactivity. In 2009, Dr. Weeks developed and formulated RelarianTM for Stephenson Products (SEO Impressions), Little Neck, NY and in 2020, he developed a line of relaxation blends for Calming Co, San Diego, CA. Dr. Weeks has recently published peer-reviewed articles on using liposomes to package and target nutraceuticals

Contents

Preface

This book presents recent advances in liposome technology and development. Liposomes are spherical phospholipid bilayer vessels with aqueous interiors that increase cellular absorption of hydrophilic and hydrophobic molecules that can range from vaccines to gene therapies to medicines and nutrients. The phospholipid bilayer structure of the liposome is known as the lamella, with some liposomes being multilamellar, containing several aqueous layers. In addition, some lamella contain multiple other vessels in what is known as multivessicular liposomes. Depending on the lamellar structure, liposomes range in size from 50 – 500 nm. While drugs and nutrients may be able to cross membranes via facilitated transport, liposomes are able to fuse with the plasma membrane and deliver the hydrophilic liposomal contents to the cell, which increases the uptake and bioavailability as well as the distribution of these liposomal contents. Further, the incorporation of aptamers or targeting ligands in the surface of the liposome that specifically binds cell-surface markers has been shown to enhance cell and tissue targeting and tissue-specific uptake of liposomal contents, including nutrients and chemotherapeutic drugs. In addition to medicine, vaccines, and gene therapies, improved nutrient uptake using liposomes has also been an important and aggressive area of research and development.

While the methodologies and procedures to create liposomes have long been established, technological improvements to incorporate multiple nutrients and drug combinations are an area of great interest and activity. For example, Pfizer Inc., New York, NY, and Moderna Inc., Cambridge, MA, developed and designed liposomes to carry the COVID-19 vaccine. Several companies, including Baxter International Inc., Skyepharma, Saint-Quentin-Fallavier, and others, have liposomal chemotherapeutic drugs as antibiotics and others on the market that are being used clinically. Further, with regard to nutrients, the LiposoMax™ technology of One Innovation Laboratories, Miami, Fl., is leading the way in developing liposomes of appropriate lamellar and vesicular structure and size to deliver combinations of nutrient molecules and improved levels of these nutrients, including vitamin C, Vitamin D3, multivitamins, glutathione and more. It is clear that there have been and continue to be significant advances in liposome technology and the delivery of medicines, nutrients, vaccines and therapies to cells and tissues. Once again, this book offers examples of liposome development and insight into the future direction of this industry.

Benjamin S. Weeks, Ph.D.
Department of Biology,
Adelphi University,
Garden City, NY, USA

Section 1

Advances in Liposome Development

Chapter 1

The Art of Liposome Surface Decoration for Targeted Drug Delivery

Justin B. Safari, Paula Maseko, Pathy B. Lokole, Galilée G. Byamungu, Simnikiwe Nogqala, Tanaka Ndongwe, Jonathan M. Mukanya, Frank Ssengooba, Gauta Gold Matlou, Xavier Siwe-Noundou, Christian I. Nkanga, Brett I. Pletschke and Rui W.M. Krause

Abstract

This chapter delves into the intricate art of surface modification of liposomes, a versatile group of nanocarriers renowned for their pivotal role in various fields, including drug delivery, diagnostics, and theranostics. Emphasising the significance of surface modification, this chapter outlines the methods employed to design liposomes with added functionalities, improved stability, and targeted delivery capabilities. The chapter offers a comprehensive overview of modern surface modification strategies from traditional conjugation chemistry to innovative bioinspired and biomimetic methods. Additionally, it meticulously examines the use of various ligands—including peptides, antibodies, proteins, and polymers—to embellish liposomal exteriors, thereby achieving enhanced biocompatibility, extended circulation times, and targeted delivery. Furthermore, the discussion extends to the implications of these surface modifications, highlighting how they influence the biological fate of liposomes, from cellular interactions to their behaviour *in vivo*. Through this exploration, the chapter aims to provide an extensive understanding of the current landscape and prospects of liposomal surface modification, fostering advancements in nanomedicine.

Keywords: liposome, targeted delivery, surface modification, drug delivery, controlled release, bioconjugation chemistry

1. Introduction

Liposomes are spherical nanostructures composed of lipid or phospholipid bilayer membranes resembling cell membranes (**Figure 1**) [1–3]. Liposomes represent the first class of nano-drug delivery systems (DDSs) to be translated into clinically

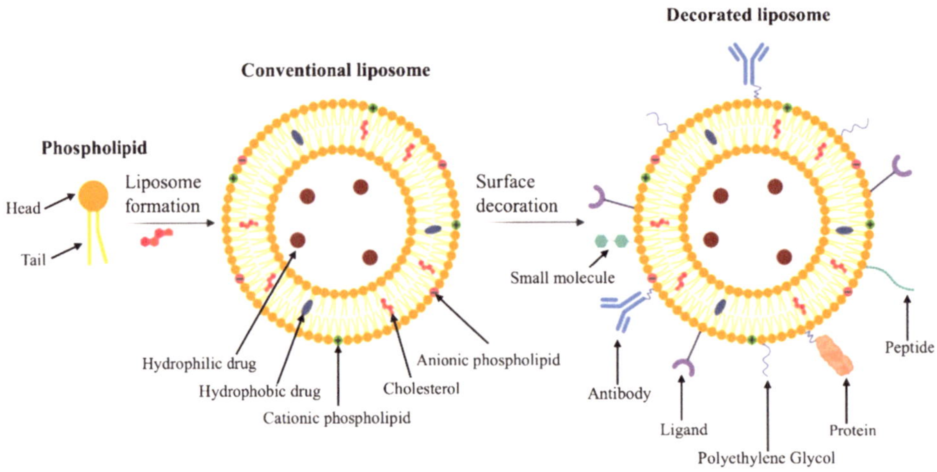

Figure 1.
Conventional and surface-modified liposomes.

successful marketed products. This success is attributed mainly to their high encapsulation capacity and diversity of payloads from small molecules to proteins while protecting from degradation, thereby facilitating controlled release and therapeutic effectiveness [4]. Additionally, liposomal DDSs possess the unique capability of encapsulating and delivering both hydrophobic and hydrophilic biomedical compounds (therapeutics and/or diagnostics) [5].

Moreover, the liposome lipid bilayer's similarity to cell membranes enhances their biocompatibility, biodegradability, flexibility, and safety [6]. Using liposomes as a DDS has enhanced treatments for various diseases by stabilising therapeutic agents, surmounting cellular and tissue absorption challenges, and enhancing the distribution of compounds to specific sites within the body. On the other hand, these same vehicles are facilitating the diagnosis of many diseases [7, 8].

Despite the advantages offered by liposomes as a DDS, conventional liposomes have shown some limitations, such as poor stability, non-specific targeting, inability to escape the immune system, and high biodistribution through the bloodstream, resulting in a variety of side effects [9, 10]. Surface modification approaches are promising pathways to tackle these limitations. Several studies have been conducted to improve liposome specificity, biocompatibility, stability, and cell penetration by functionalising conventional liposomes with chemicals that can increase the performance of liposomes [11].

These decorators/modifier agents can be bound to the surface of liposomes via non-covalent interactions, such as van der Waals forces, H-bonds, electrostatic interactions, or covalent bond formation, to mention a few [9, 12, 13]. As an example, covalent bonding can improve *in vitro* and *in vivo* stability of liposomes by allowing rapid fusion to form lipid films and sustain fluidity and increasing the required energy or bioreagents for bond breaking compared to non-covalent interactions (typically, this is a 10-fold improvement over electrostatic interaction) [14, 15].

In this chapter, we explore various strategies and technologies employed in liposome surface modification, highlighting their impact on improving liposomal nanocarriers' performance and therapeutic efficacy in biomedical applications.

2. Surface modification techniques

2.1 Overview of conjugation chemistry methods

The functionality of liposomes, which are often used as carriers in drug delivery, diagnostics, and other biomedical applications, can be improved by conjugation chemistry techniques [16]. Two molecules can be joined using several methods, crosslinking, utilising the reactive groups present on the molecules, or activating agents that generate an intermediate reactive group on one component for coupling. These reactive groups then interact with specific functional groups on the other molecules, facilitating conjugation. Liposome conjugates are created by modifying either the liposome, the molecule to be attached, or both. The use of crosslinkers or reactive groups to facilitate coupling is common [17–19]. In the following paragraphs is a detailed overview of the traditional conjugation chemistry methods focusing on their application to liposomes.

2.1.1 Amine-reactive chemistry with liposomes

Amine-reactive chemistry is frequently used to conjugate compounds with primary amine groups, including proteins and amino acids. Through the use of reagents, such as N-hydroxysuccinimide (NHS) esters, proteins [20], peptides, succinimidyl 3-[2-pyridyldithio)propionate (SPDP), succinimidyl-4-[*N*-maleimidophenyl]butyrate (SMPB), or succinimidyl-4-[*N*-maleimidomethyl]cyclohexane-1-carboxylate (SMCC) co-linkers can be attached to the liposome. This is often done to improve or enable targeted drug delivery [18, 21].

The mechanism of amine-reactive chemistry with liposomes typically involves the activation of amine groups ($-NH_2$) on the liposome surface, followed by their reaction with reactive moieties on molecules to be attached. Initially, amine groups ($-NH_2$) are activated using reagents, such as NHS esters or isothiocyanates (**Figure 2A**). These reagents form an activated intermediate, such as succinimides or thiourea intermediates [16, 22]. These unsaturated electrophilic species then react with nucleophilic groups on the corresponding attaching molecules [20].

For instance, primary amines ($-NH_2$) or hydroxyl groups ($-OH$) on molecules, such as proteins, peptides, or targeting ligands, can react with the activated intermediates on the liposomes, resulting in the formation of stable covalent bonds [22]. This process allows for precise and controlled functionalisation of liposomes with a wide range of molecules for specific biomedical applications [18].

2.1.2 Carbodiimide-mediated coupling with liposomes

Carbodiimide chemistry provides yet another method for conjugating molecules to liposomes through the activation of carboxyl groups. Using popular carbodiimide reagents like 1-ethyl-3-(3-dimethylaminopropyl) carbodiimide (EDC) and primary amines, including proteins and peptides, liposomes can be functionalised for various applications, including targeted drug delivery or vaccine development [18].

The carbodiimide reagent activates selected carboxyl groups on the liposome surface, resulting in the formation of an active intermediate, typically an *O*-acylisourea intermediate, as shown in **Figure 2B** [22]. The activated carboxyl group then reacts

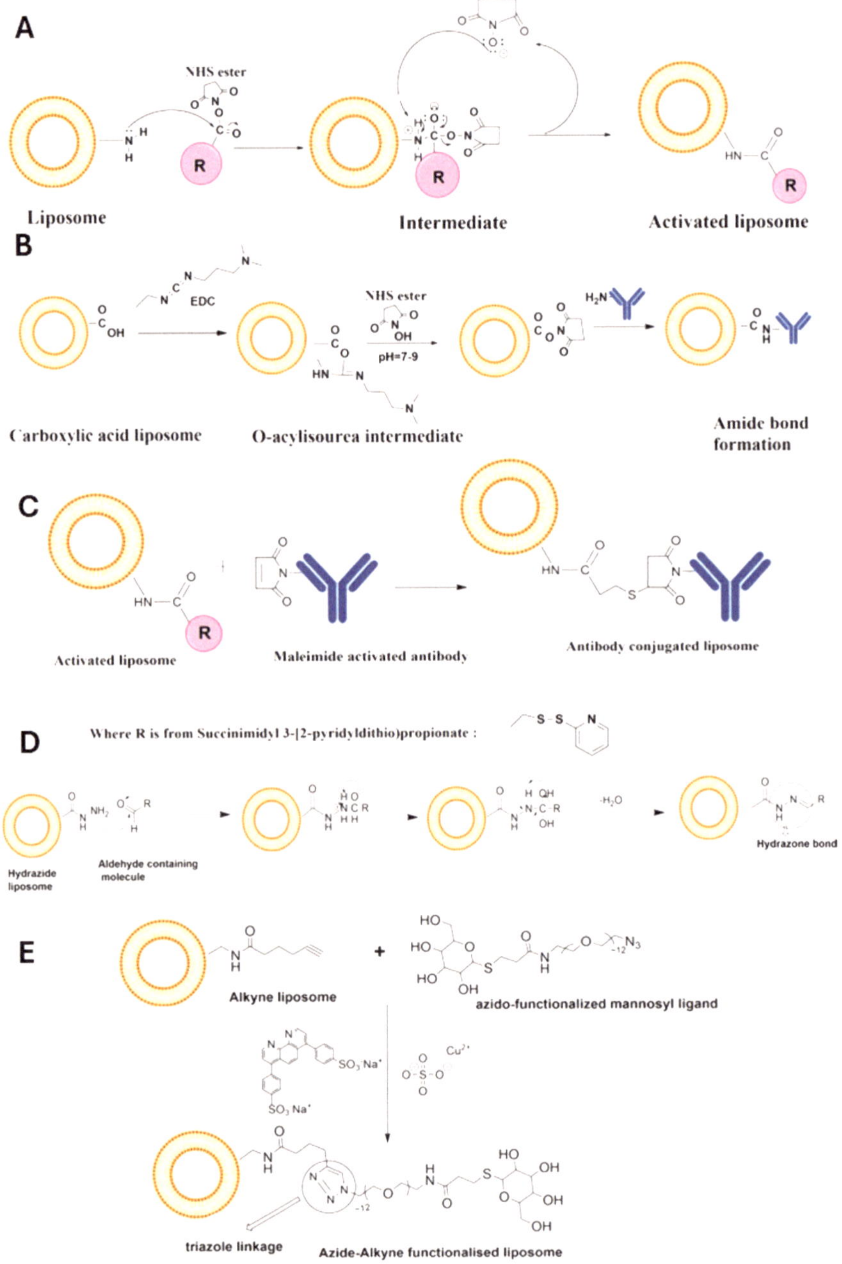

Figure 2.
Schematic illustrations of chemical reaction used for liposomes surface functionalisation. (A) Liposome activation using NHS esters, (B) carbodiimide-mediated coupling with liposomes, (C) thiol-reactive chemistry with activated liposomes and maleimide-activated antibodies, (D) hydrazone bond formation, and (E) click chemistry.

with a primary amine on the molecule intended for attachment resulting in the formation of an amide bond between the liposome and the newly attached molecule. Any unreacted *O*-acylisourea intermediates are usually hydrolysed under the reaction conditions, leading to the formation of urea derivatives and releasing the active carbodiimide byproduct [16].

Overall, carbodiimide-mediated coupling with liposomes involves the activation of carboxyl groups on the liposome surface using carbodiimide reagents such as EDC, followed by their reaction with primary amine groups on molecules to be attached. This process enables the formation of stable covalent amide bonds between the liposome and the attached molecules [22].

2.1.3 Thiol-reactive chemistry with liposomes

Thiol-reactive chemistry offers a versatile orthogonal approach to amine activation for modifying liposomes. By introducing thiol moieties or using thiol-reactive reagents like maleimides, liposome surfaces can be subsequently conjugated reactions or reacted with various electrophiles. Common modifications include polyethylene glycol (PEG) chains for enhanced stability, fluorescent dyes for imaging, or targeting ligands for specific cell recognition and uptake (**Figure 2C**) [22, 23]. Electrophilic agents, such as maleimides or pyridyl disulphides, are commonly used for thiol-reactive chemistry, typically in a facile manner such as activation by reduction or deprotonation [16, 23].

The reaction proceeds via a nucleophilic attack of the thiol group on the electrophilic centre of the agent, resulting in the formation of a stable covalent bond such as a thioether, providing a robust linkage between the thiol-containing molecules and the liposome surface [18, 23].

Thiol-containing molecules, such as peptides, proteins, and small organic ligands, can be conjugated to the liposome surface in this manner, enabling the conjugation of thiol-containing moieties for various biomedical applications [12, 19, 21, 24].

2.1.4 Click chemistry with liposomes

Click chemistry offers a highly selective and efficient method for conjugating molecules to liposomes with minimal side reactions. Liposomes can be functionalised with azide or alkyne groups on their surface through the incorporation of lipids bearing azide or alkyne moieties during liposome preparation or through post-functionalisation methods [17]. These functional groups react with their complementary reactive partners using click chemistry conditions, enabling precise control over the attachment of molecules and facilitating applications, such as imaging, biosensing, or targeted drug delivery. Copper-catalysed azide-alkyne cycloaddition (CuAAC) and strain-promoted azide-alkyne cycloaddition (SPAAC) are the most used methods for conjugating molecules to liposomes through click chemistry because of their high efficiency and selectivity under mild conditions [16, 18].

In CuAAC, copper(I) ion is used as a catalyst for the reaction between the azide and alkyne groups. Cu^+ is generated from Cu(II) precursor in the presence of reducing agents, such as sodium ascorbate or tris(2-carboxyethyl)phosphine (TCEP) [25]. The Cu(I) catalyst activates the terminal alkyne group by coordinating with it. The activated alkyne then reacts with the azide group through a 1,3-dipolar cycloaddition reaction, forming a stable triazole linkage resulting in the formation of a covalent triazole bond between the liposome and the attached molecule [26]. After the reaction, Cu^+ is oxidised back to Cu(II), completing the catalytic cycle. This process allows for the recycling of the copper catalyst and facilitates the efficient coupling of azide and alkyne groups on liposomes [25, 26].

On the other hand, SPAAC does not require a metal catalyst but relies on the inherent reactivity of strained cyclooctyne derivatives, which can react selectively with azide groups in the absence of copper catalysts [25]. The strained cyclooctyne derivative undergoes a spontaneous reaction with the azide group on the liposome surface, forming a stable triazole linkage, as with CuAAC, resulting in the formation of a covalent triazole bond between the liposome and the attached molecule.

Click chemistry with liposomes involves the selective and efficient conjugation of molecules through the reaction between azide and alkyne functional groups, either catalysed by copper in CuAAC or promoted by strain in SPAAC. This versatile approach allows for the precise functionalisation of liposomes under mild conditions, making it suitable for various biomedical applications.

2.1.5 Hydrazone formation with liposomes

Hydrazone formation involves the selective and reversible reaction between hydrazide groups on the liposome surface and aldehyde or ketone groups on molecules to be attached. This dynamic covalent bond formation strategy allows for the precise functionalisation of liposomes with a wide range of molecules. By reacting hydrazide-functionalised liposomes with aldehyde- or ketone-containing molecules, liposomes can be formulated for applications, such as triggered drug release in response to specific stimuli or the preparation of smart liposomal systems for controlled drug delivery [18]. Hydrazide groups can be introduced through various methods, such as using hydrazide-modified lipids during liposome preparation or several post-functionalisation techniques. These hydrazide groups serve as reactive sites for subsequent conjugation reactions [16].

Aldehyde (−CHO) or ketone (C=O) moieties can be introduced through oxidation or other chemical modifications, and these moieties then react with hydrazide groups on the liposome surface via a condensation reaction, forming a hydrazone bond. Moreover, the stability of the hydrazone bonds can be further enhanced by additional modifications or changing the reaction conditions, such as pH and temperature, to ensure efficient coupling and minimise hydrolysis [16]. Incorporating liposomes into traditional conjugation chemistry methods provides a versatile platform for the development of functionalised liposomal systems with tailored properties and functionalities. These conjugation strategies enable the design of advanced liposomal delivery systems with improved stability, specificity, and therapeutic efficacy, paving the way for innovative biomedical applications. These methods provide versatile tools for the functionalisation of biomolecules, nanoparticles, polymers, and surfaces, enabling the development of advanced materials. They offer precise control over the attachment of molecules, allowing for the design of tailored conjugates with desired properties and functionalities.

2.2 Bioinspired and biomimetic approaches for surface decoration

Apart from traditional chemical transformations, biological systems present interesting innate characteristics that continue to inspire the engineering of drug delivery systems such as liposomes. Bioengineering, bioinspired, and biomimetic processes have emerged as schemes to overcome the hurdles of synthetic vesicle limitations [27, 28]. Furthermore, understanding the cell-based interactions in biological systems enables scientists to mimic core cellular or viral aspects in synthetic systems, for instance, by using similar particle size and shape. Furthermore, surface features can also be exploited to mimic biological and physicochemical properties of biological system surfaces that improve specific targeting, membrane permeability, and stealth properties of delivery systems [29, 30].

Several approaches using supramolecular or conventional chemistry are exploited to decorate the surface of liposomes by mimicking biological structures. Liposomes can be functionalised with fusogenic molecules, bio-orthogonal reactions, cell-molecule engineering, etc.

2.2.1 Supramolecular decoration of liposomes

Supramolecular chemistry could be described as the organised chemistry of non-covalent bonds [31]. Non-covalent bond formation at the surface of liposomes is a way to mimic cell membranes, and this can be done using different approaches, such as physical adsorption and electrostatic interactions. These approaches have been used to good effect to mitigate foreign-body reactions [15, 32, 33].

The development of "supramolecular liposomes" featuring cytosine moieties at the surface has been reported. This was achieved by the assembly of single-nucleolipid molecules consisting of a hydrophilic deoxycytidine-3′-phosphate linked to a glycerol dipalmitic lipid chain (diC16dC) and has been applied for silver sensing applications. The coordination of Ag^{+} ions by the nucleotide moiety enhanced membrane rigidity, which increased the fluorescence of a common reporter, Thioflavin T [34].

Another study reported the formulation of a virus-mimicking pH-sensitive delivery platform by surface functionalisation of the self-assembled liposomes (consisting of L-α-phosphatidylcholine from egg yolk and cholesterol) with influenza viral peptide-mimicking anionic pseudopeptide polymers. Hydrophobic interactions were crucial in the impact of the affinity of polymers for phospholipid vesicles for non-covalent polymer-liposomes complex. The result indicated no leakage at neutral pH, while the complete release was observed at endosomal acidic pH. Intracellular delivery of doxorubicin, an anticancer drug, was also investigated using membrane-impermeable calcein. Doxorubicin (DOX) fluorescence was observed an hour after incubation into HeLa cells, confirming this virus-mimicking system's high efficiency in doxorubicin delivery into cancerous cells [35].

2.2.2 Liposomes functionalisation with fusogenic molecules

Taking advantage of the similarity of the structure of liposomes with cell membranes, the fusion between liposomes and cell membranes is a promising approach for the direct delivery of payloads in eukaryotic cells [36, 37]. This complex process can be facilitated using fusogenic stimuli, such as reconstituted proteins, fusion peptides, electric pulses, etc. [38].

Multiple approaches can be used in fusogenic liposomal formulation, ranging from the incorporation of special lipids to improve the fluidity of the membrane and enable the promotion of membrane destabilisation to Sendai virus inactivation [39].

Additionally, fusogenic peptides can also be displayed on liposome surfaces through the conjugation methods described above. A study by Kakudo et al. assumed that the disposition of a fusogenic peptide (glutamic acid-alanine-leucine-alanine (GALA)) on the surface of liposomes was necessary to ensure an efficient release and diffusion of rhodamines into the cytosol using transferrin as a cell recognition agent for cell entry and GALA as fusion-induction device. The results suggested that ~30% of the liposomes were modified on the surface with cholesterol-GALA fused at pH 5.0, while no fusion promotion was observed with liposomes that were unmodified on the surface. A cytosolic release of sulforhodamine B was observed in almost all cells at 24 hours when transferrin attached to liposomes was coated with cholesterol-GALA incubated in the same study [40]. The challenge remains to incorporate peptides on the surface of liposomes in a stable manner by preventing peptide insertion into the membrane [41]. Covalent conjugation of the head group to an enzyme-susceptible peptide could mask its fusogenic potential until specific enzyme cleavage [42].

Motion et al. placed phosphate groups into a Human Immunodeficiency Virus (HIV) gp41 N-terminus fusion peptide to accomplish membrane insertion and improve peptide polarity character. The presence of phosphate groups in fusion peptides masked their hydrophobic character and inactivated their fusogenic properties. Phosphatases were used as the enzyme for cleavage, resulting in regeneration of membrane-destabilisation properties and restitution of fusogenic activity. Phosphatase-triggered fusogenic liposomes could be a promising pathway for mediating the cytosolic controlled release of therapeutics against diseases in which phosphatases are overexpressed such as cancer [43].

2.2.3 Bio-orthogonal reactions

Bio-orthogonal chemistry allows the processing of synthetic chemical reactions in living biological systems without jeopardising biochemical processes or affecting the functioning of biomolecules [44, 45]. Several bio-orthogonal reactions have been used in the surface modification of liposomes, including alkyne-azide cycloaddition, Diels-Alder, Michael addition, and oxime formation [46, 47].

Self-assembled materials such as liposomes can be pre-modified or post-modified. Post-modification methods have been proposed as an alternative to pre-modification methods to avoid chemical modification of component and side reactions with sensitive cargos. Those techniques involve bio-orthogonal approaches and specific functionalisation of vesicle surface after vesicle formulations [48, 49]. The chemical reactions employed for this approach are discussed in Section 2.1 above.

For example, Feldborg et al. performed chemoselective and bio-orthogonal reactions for conjugation of a neuroendocrine tumour targeting peptide (TATE) both in solution and on liposomes surface. Oxime formation, CAA copper-catalysed (CuCAA), strain-promoted CAA copper-free (SPACC), and Michael addition reactions were all used for various conjugation reactions. Although all reactions carried out on the surface of functionalised vesicles were slower than the reaction in solution, SPACC reactions were faster after applying post-functionalisation with a reaction yield value of 84% after 3 days (compared to 86% after 50 days in solution). Otherwise, the surface conjugation reaction showed highly reproducible properties [50]. The proposed scheme is a promising approach for formulating targeted tools for theragnostic applications (**Figure 3**).

2.2.4 Surface functionalisation with cell molecules

Liposomes have been described as a simplified model of cell membranes with no ornaments, but increasingly modified liposomes have been functionalised with membrane proteins and carbohydrates for multiple purposes, including biosensing and immune system activation [51, 52]. Those molecules have inspired liposomes coated with antibodies and peptides for active targeting of cells bearing the corresponding antigens or receptors [35, 53].

Liposome features have been developed as an alternative model to animal experiments by mimicking cells such as nerve and hepatic cells [54].

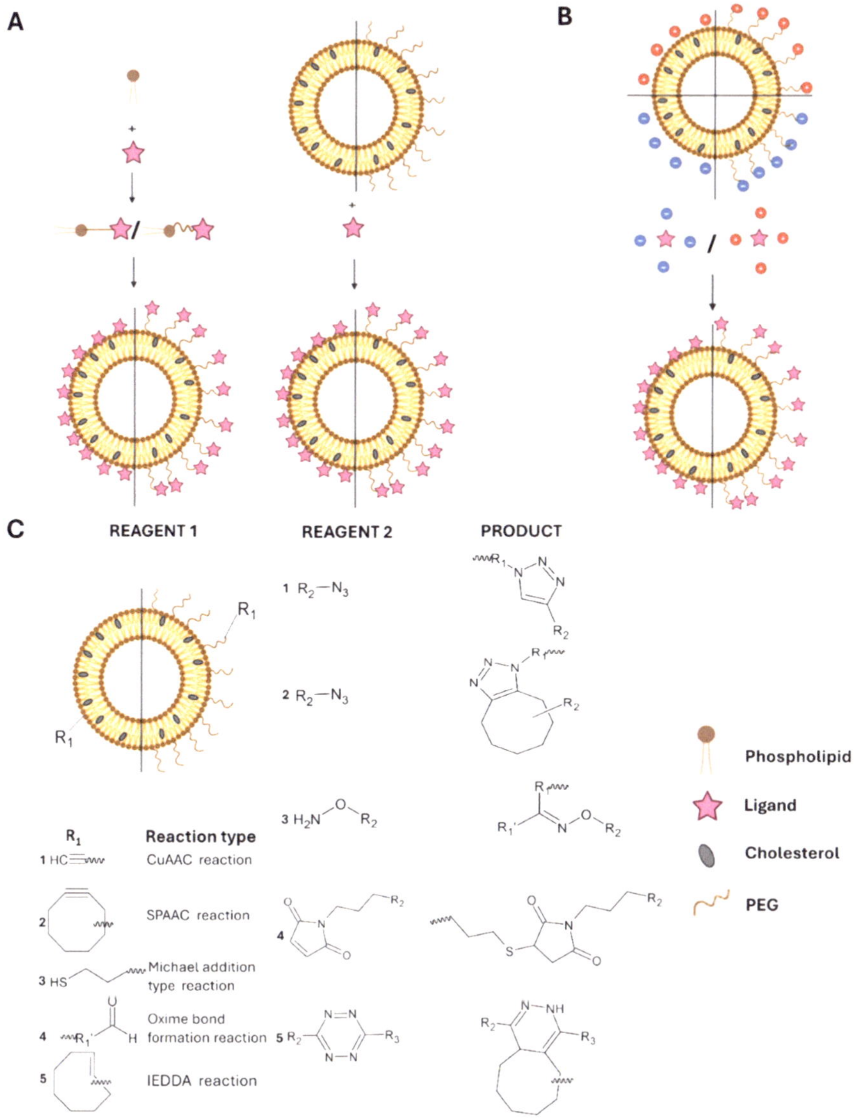

Figure 3.
Schematic illustrations of bioinspired and biomimetic approaches used for liposome surface functionalisation. (A) Pre-modification (left) and post-modification (right) methods for conventional and PEGylated liposome functionalisation. (B) Non-covalent incorporation of ligand on the surface of cationic or anionic conventional or PEGylated liposomes. (C) Summary of bio-orthogonal reactions used in liposome surface modifications. CuAAC: alkyne-azide cycloaddition copper-catalysed, SPAAC: strain-promoted alkyne-azide cycloaddition IEDDA: inverse electron-demand Diels-Alder. BioRender (online) and ChemSketch 2021.2.1 version C35E41 were, respectively, used to create the structures in the two panels.

Ganglioside-functionalised liposomes have proven their efficiency in cholera toxin detection and various physiological activity investigations of botulinum neurotoxins [55, 56]. In the same concept, hepatic lysosomes were developed from non-covalent attachment of horse-spleen ferritin to phospholipid suspended in agarose to mimic liver iron overload in transfusion-dependent anaemias [57].

Glyco-functional liposomes have been reported to be a suitable platform for carbohydrate receptor-mediated targeting. In this concept, glycopolymer augmented liposomes—synthesised by the addition of a copolymer of glycopolymer-co-cholesterol during the rehydration of the lipid this film—have proven their efficiency in the expression of macrophage mannose receptor (CD206) and macrophage galactose-type lectin (CD301) [35, 58]. In addition, liposome surface glycan engineering is a promising pathway in endogenous immune response activation for diseased cells targeting and elimination [59]. Rhamnose-functionalised liposomes have been developed for immune targeting leading to tumour cell destruction. In fact, decoration of liposomes with rhamnose resulted in the promotion of anti-rhamnose recruitment, thereby activating immune responses for complement-mediated cell killing [52].

3. Types of ligands for liposome decorations

In recent years, the use of various types of ligands that play an indispensable role in the surface modification of liposomes has risen exponentially. Ligands are essential in improving drug delivery systems and have greatly influenced the development of precision medicines, as depicted in **Table 1**.

Among the leading types of ligands in targeted delivery are small molecules (< 1000 Daltons) and molecules with high binding and selectivity traits in both the diagnosis and treatment of diseases [66–68]. In this respect, factors, such as geometry and chemical complementarity, are carefully considered to improve the precision aspects of the target. In drug delivery, liposomes that are functionalised through surface modification by the use of small molecules have proven to be potentially effective [69]. Examples of small molecules, including affibodies, carbohydrates, and folates, have been widely reported as liposomal surface modifications [70, 71].

Using peptides as ligands has also drawn much attention due in part to the availability of facile synthetic methods, the ubiquitous utility of peptides, the chemical versatility, as well as their high affinity for targeted receptors [72]. Consequently, the site specificity of liposomes can be significantly increased through functionalisation with these ligands [24, 73]. In this regard, surface modification can be achieved through covalent and/or electrostatic functionalisation as attached peptide ligands [74]. For example, peptides such as Peptide T7, sequence HAIYPRH are frequently used in the targeting of transferrin receptors, which are often expressed in cancer cells [75]. Another peptide ligand called GE11, also known as YHWYGYTPQNVI, demonstrated a strong affinity for epidermal growth factor receptor (EGFR) and was subsequently employed for liposomal conjugation to harness its EGFR-targeting properties. Notably, the peptide sequence GE11 has proven to be a valuable tool in exploiting the capabilities of EGFR targeting [60]. Another study by Carvalheiro and workers evaluated the utilisation of antagonist G, a peptide-targeting agent with the ability to hinder the function of various neuropeptides, to enhance the specificity of targeting and internalisation of liposomal formulations, specifically long-circulating liposomes [76].

Similarly, antibodies are a class of target ligands that have proven to be indispensable in surface modification studies [67, 77]. Of interest is the use of antibodies in chemotherapy to multidrug resistance, improve drug internalisation, reduce circulation time, improve site specificity of drugs, and minimise the toxicity of drugs. Immunoliposomes have been used to improve targeted delivery and enhance the

Liposome composition	Ligand type	Role	Ref.
PEGylated liposomes (PEG-LIP), HSPC/ cholesterol/1,2-disteroyl-sn-glycero-3-phosphoethanolamine-*N*-[carboxy(polyethylene glycol)2000, NHS ester] (sodium salt)	DT7-Cys (D(HRPYIAHC)) and LT7-Cys (L(CHAIYPRH))	Improved targeting and efficacy	[60]
Egg yolk phosphatidylcholine (EPC), cholesterol (Chol) and 1,2-disteroyl-sn-glycero-3-phosphoethanolamine-*N*-[methoxy(polyethylene glycol)-2000] (DSPE PEG 2000)	polyethylene glycol (PEG) and polyvinyl alcohol (PVA)	Lower affinity to the liver and long residence in the systemic circulation	[61]
Cholesterol	Chitosan oligosaccharide (CSO)	Increased affinity for tumour cells	[62]
1,2-dioleoyl-3-trimethylammonium-propane, 1,2-distearoyl-*sn*-glycero-3-phosphocholine, and 1,2-distearoyl-*sn*-glycero-3-phospho-(1′-rac-glycerol) (DSPG)	Polyethylene glycol	Affects the surface binding and subsequent uptake of liposomes in Caco-2 cells	[63]
1,2-disteroyl-sn-glycero-3-phosphoethanolamine-*N*-[succinimidyl(polyethylene glycol)] (DSPE-PEG-NHS), *N*-(Succinimidyloxy-glutaryl)-L-α-phosphatidylethanolamine, Distearoyl (DSPE-NHS) and 1,2-distearoyl-sn-glycero-3-phosphocholine (DSPC) Cholesterol	Protein A-R28	Improved selective targeting	[64]
Cholesterol, 1,2-distearoyl-sn-glycero-3-phosphoethanolamine-*N*-[methoxy(polyethylene glycol)-2000] (ammonium salt) (DSPE-PEG2000), and 1,2-distearoyl-sn-glycero-3-phosphoethanolamine-*N*-[dibenzocyclooctyl(polyethylene glycol)-2000] (DSPE-PEG2000-DBCO)	Antibody FLAG tag (peptide sequence DYKDDDDK)	Improved selective brain targeting and uptake	[65]

Table 1.
Some of the reported studies on the surface modification of liposomes.

pharmacological activity of drugs, especially in cancer therapy [64]. In the modification of liposome surfaces, the predominant approach involves utilising fragments derived from small antibodies that contain the complementarity-determining regions (CDRs) [78]. Ye et al. demonstrated using receptor-mediated transcytosis (RMT) liposomal system as a promising delivery system in targeting the blood-brain barrier [65].

Protein use is an extension of this kind of surface modification of liposomes, which has also significantly increased over the years [69]. Abdelrehim and colleagues investigated the role of tail-anchored (TA) membrane protein (cytochrome b5) in the surface modification of liposomes loaded with carboxyfluorescein. The study results revealed that using TA proteins improved liposomal drug delivery. In addition, proteins have been cited to play a crucial role in anchoring liposomes, improving polarity and affinity to the binding site [79–81].

Synthetic- and biopolymers are yet another important class of modifying ligands that have played a prominent role in stabilising liposomes and increasing drug activity in some cases. Techniques including grafting or adsorption of both synthetic and natural polymers are among the leading approaches used in liposomal surface modification [63, 82]. To enhance the delivery of paclitaxel in lung cancer treatment,

Miao et al. used chitosan oligosaccharide (CSO) for the surface modification of liposomes. The authors prepared CSO-conjugated Pluronic P123 polymers in different amounts and reported that these CSO modifications showed promising results for the development of nanomedicines in the treatment of lung cancer (CSO has good mucoadhesion) [62]. Shehata et al. studied the biodistribution profile of liposomes that were surface-modified using PEG and polyvinyl alcohol (PVA). The outcome of the study revealed increased systemic circulation and lower drug distribution. Also, a significant number of studies have revealed the broad application of polymers in the surface modification of liposomes [61]. **Table 1** summarises some of the reported applications of diverse ligands in surface-modified liposomes.

4. Functional enhancements and biological impacts

4.1 Roles in improving biocompatibility and circulation time

Liposomes must be safe to use as drug delivery systems, so often, biocompatible lipids are employed to this end in formulations [83, 84]. PEG, a biodegradable polymer, is widely used in modified liposomes to improve biocompatibility and enhance anticancer drug permeability and retention effect in solid tumours [77]. This successful application has made liposomes increasingly used as an efficient drug delivery system. However, some PEGylated liposomes (PEG-LIP) are recognised by anti-PEG immunoglobulins (IgM) after intravenous administration, which can cause accelerated blood clearance, although this mainly occurs with methoxy (-OCH_3)-PEG. The effect of anti-PEG IgM on different functional groups of PEG-coated liposomes, such as carboxyl-, hydroxyl-, amino-, and methoxy-PEG, has been studied, and it has been determined that hydroxy-PEG-coated liposomes pose low risk of recognition by anti-PEG IgM [85]. Additionally, an investigation into the placement of D-α-tocopheryl PEG 1000 succinate (TPGS) in the decoration of liposome surfaces showed that TPGS-coated liposomes can improve the pharmacokinetic properties of liposomal delivery of drugs for oral and ocular routes as well as vaccines. Most pertinently, TPGS helps to enhance the circulation half-time of conventional liposomes [11].

The enhancement of the circulation time of liposomes depends on their stability. Hollmann et al. focused on the stabilisation of liposomes formulated by mixing soybean lecithin, cholesterol, and stearylamine coated with natural proteins from *Lactobacilli* (*Lactobacillus brevis* and *Lactobacillus kefir*). Calcein was used to study the encapsulation capacity and stability of liposomes as functions of pH (2.5, 4, and 7) and time (30 and 60 minutes) in coated and uncoated liposomes. Results showed that all liposomes had high encapsulation efficiency at pH 7, with 91.2, 89.5, and 77.3% for liposomes coated with proteins from *L. brevis*, *L. kefir*, and the control, respectively, after 30 minutes of incubation. These results decreased after 60 minutes of incubation to 82.0, 86.5, and 51.2% for the same set of coated liposomes [86]. In their study, Tran et al. emphasised the usefulness of pectin-coated liposomes by extracting pectin from *Tithonia diversifolia*, which was mixed with phosphatidylcholine and Tween 80 to produce liposomes using the thin film hydration method. The results highlighted the advantages of coating liposomes with pectin, as the zeta potential (ZP) was decreased from −17.5 to −45.7 mV, allowing the stability of liposomes. These pectin-coated liposomes showed a slight increase in the rate of the encapsulated anticancer (tagitinine C), with an encapsulation efficiency of 85.7 and 89.8% for uncoated and pectin-coated liposomes, respectively. Tagitinine C, encapsulated in pectin-coated liposomes, showed more cytotoxic effects than the uncoated one [87].

Some drugs particularly those with narrow therapeutic indexes, including anti-cancer agents, are associated with severe adverse effects [88]. Using liposomes to deliver such narrow therapeutic index drugs has decreased their toxicity by targeting specific receptors. Polysaccharides are often used to decorate liposomes to allow them to target cancer cells specifically. Being of biological origin, they evince excellent biocompatibility and non-immunogenic properties [89]. Of the polysaccharides, hyaluronic acid (HA) is widely used as a biomarker of cancer cells as it possesses an interesting affinity with CD44 overexpressed in solid tumours, such as breast, colon, lung, and ovarian tumours [90]. In a study by Ding et al., liposomes decorated with chitosan oligosaccharides combined with photochlor were used to target CD44, highly expressed in triple-negative breast cancer, and to irradiate tumour sites with a 660-nm light-mediated photosensitiser for photodynamic therapy and evofosfamide, a prodrug. This decorated liposome proved to be highly biocompatible, has targeting capability *in vitro,* as well as better imaging capability than non-decorated liposomes [91]. In another study, Alavi et al. demonstrated that chitosan-coated liposomes can be more efficient than uncoated liposomes by increasing the zeta potential of liposomes from −1 to +21 mV, leading to size and shape control, interaction with the environment, enhancing the specificity of sirolimus encapsulated in liposomes to attenuate vascular restenosis, and better stability against Triton X-100, a non-ionic surfactant and detergent able to lyse the liposome membrane [92]. HIV treatment has been improved by using coated liposomes to enhance the specific targeting of zidovudine (ZDV) on lymph nodes and spleen. To this end, Kaur et al. formulated liposomes with stearylamine and dicetylphosphate by thin film hydration. Mannose was used as the coating agent of those liposomes, and 6-carboxyfluorescein (6-CF), used as a fluorescent biomarker, was encapsulated into coated and uncoated liposomes. Their results showed that coated liposomes significantly reduced ($p < 0.05$) the drug release profile of ZDV compared to the uncoated one. When studied via fluorescent microscopy, the extracted spleen and lymph of albino rats' nodes presented a uniform distribution of 6-CF to the targeted cells and tissues [93].

Overall, despite their advantages as drug delivery systems, liposomes need to be decorated with biocompatible compounds to remain non-toxic and avoid immune system capture. Once in the bloodstream, the liposomes should aim at a specific location and selectively deliver therapeutic and/or diagnostic products.

4.2 Achieving specific targeted capabilities through surface decorations

Liposome surface decoration plays a crucial role in tailoring liposomes for specific functions. Re-designing and engineering the surface of liposomes could significantly improve targeted capabilities towards specific tissues, cells, organs, etc. Previous studies, particularly Moulahoum et al., have demonstrated that modifying the surface of liposomes can overcome some of the limitations associated with undecorated liposomes [74].

Many approaches have been adopted to use liposomes to target specific sites, including passive/spontaneous liposomal targeting and active liposomal targeting sites [69]. This section discusses some organ/tissue-specific target delivery realised by surface modification.

Magnetic fluid-loaded liposomes (MFLs) with optimised magnetic reactivity have been developed by trapping superparamagnetic maghemite nanocrystals in submicron phospholipid vesicles labelled with PEG and rhodamine. These magnetoliposomes are an effective tool for the selective magnetic targeting of malignant tumours located in the brain, and non-invasive magnetic resonance imaging (MRI) tracking through intravascular administration. This selective and precise MRI-traceable targeting holds great promise for therapeutic applications, as healthy brain tissue can be expected to be spared during treatment with deleterious anticancer drugs carried by magnetically guided MFLs [94].

Surface modification of liposomes is considered to be a promising method to improve and facilitate the intestinal absorption of carried molecules [13]. The modifiers/decorators can bind to liposomal surfaces by van der Waals forces and electrostatic interaction; they also can conjugate with liposomal phospholipids through a catalyst to generate hydrogen and covalent bonds.

Zhang et al. used biotin-DSPE (distearoyl phosphatidylethanolamine) to modify surface liposomes to deliver insulin, which improved insulin absorption efficiency through intestinal biotin receptor-mediated endocytosis. The results showed that the relative pharmacological bioavailability of biotin-modified liposomes was 8.95% higher than that of conventional liposomes [13]. T7 peptides with a terminal cysteine were used to decorate liposomes loaded with quercetin (QR), which enhanced the therapeutic efficacy of QR in targeting the treatment of lung cancer [95].

In another study, surface liposomes were modified using biotin and fructose to create dual targeting functions, targeting MCF-7 cells and 4 T1 cells to treat breast cancer. The results showed that the content of modified liposome ingested by MCF-7 cells was 3.11 times that of conventional liposome, and the content of modified liposome ingested by 4 T1 cells was 3.27 times that of conventional liposome. The modified liposomes showed a strong inhibitory effect and apoptosis rate of cancer cells [13]. Shahin and co-workers formulated a targeted liposome doxorubicin (DOX) using an engineered breast tumour targeting peptide ligand and demonstrated that this surface modification with engineered p18-4 peptide at an optimum density can improve the antitumour efficacy and selectivity of liposomal DOX in breast cancer [96].

Lymphoid tissues have also been targeted using surface-modified liposomes as a drug delivery system. This avoids the non-selective action of anticancer drugs, which can lead to significant toxicity in many rapidly dividing normal cells, including bone marrow cells and hair follicles. ZDV was specifically delivered to the lymphatics using surface-modified liposomes by the incorporation of charges (positive or negative) and a specific ligand (mannose) to improve localisation in the lymph nodes and spleen [93].

4.3 Improving stability and functionality for specific applications

Despite the positive properties of liposomes, they are inherently classified as moderately unstable colloidal systems. Conventional liposomes are quickly eliminated, making them inefficient carriers [97]. Generally, to assess the stability of a liposome, one should examine several specific parameters, including chemical and physical stability, preservation of size and structure, maintenance of encapsulated drugs, and the effects of biological fluids on liposomal properties. The Food and Drug Administration (FDA) stipulates that liposomes must maintain stability for a minimum of 2 years to qualify as a liposomal drug product [98]. This requirement underscores the importance of robust stability testing to ensure the effectiveness and reliability of liposomal formulations over extended periods.

Choosing an appropriate liposomal formulation involves selecting the right combination of liposome composition, functionalisation, and targeting approach. Liposomes can exhibit diverse functionalities based on differences in head groups, aliphatic chains, and the saturation levels of fatty acids [99]. In their typical state, liposomes often exhibit limited stability due to their delicate phospholipid membranes. Modifications or coatings are applied to the phospholipid bilayers using various materials to improve their stability. This approach has been implemented both in pharmaceutical and food applications of liposomes.

The selection of phospholipids, including their head group and chain length, as well as the proportion of liposomal components, is critical in determining the safety, stability, and efficacy of liposomes [100]. Apart from phospholipids, there are other components, including but not limited to glycols like propylene glycol and PEG, cholesterol (Chol), and even polymers such as chitosan, which can further improve the stability of liposomes [101]. Being hydrophobic, Chol induces tight packing of the phospholipids and hinders interactions within the lipid chains by inserting itself between them, facilitating the stabilisation of the liposome membrane (**Figure 4A**). In the absence of Chol, liposomes can interact with different proteins (macroglobulin, transferrin, high-density lipoproteins, and albumin) destabilising the structure of the liposomal membrane and decreasing their performance as drug carriers [102–104].

The structure and attributes of phospholipids significantly impact the properties of the resulting liposome. The stability of the liposome plays a crucial role in regulating the controlled release of active compounds in the bloodstream (injectables) and intestinal mucosa (oral administration). Physical factors, such as storage temperature and light, along with chemical factors like lipid peroxidation and pH fluctuations, are influential in determining the stability of the liposome [105].

One of the major problems of liposomes is their rapid elimination by the organism's defence mechanisms (such as phagocytic uptake). There are different ways to alleviate this problem including the application of a protective coating, the addition of cryoprotectants before lyophilisation, changing the liposomal membrane structure, and the use of surfactants [106–109]. The stability of these liposomes can be affected by the gastric juice's low pH, resulting in their degradation and release of the loaded molecules in the gastrointestinal tract [110].

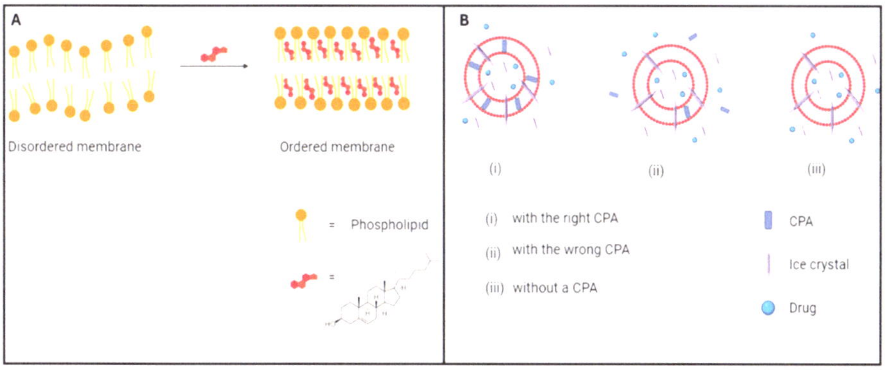

Figure 4.
(A) Stabilising liposome lipid bilayer with cholesterol, (B) stabilising liposome with a cryoprotectant (CPA) where (i) with the right CPA, (ii) with the wrong CPA, and (iii) without a CPA.

Another concern regarding liposomes is their long-term storage stability and preservation. To address this issue, cryoprotectants, such as dimethyl sulphoxide (DSMO), glycerol, sugars, disaccharides, and polyampholytes, have been employed to modify their surface during lyophilisation. Cryoprotectants, also known as CPAs, play a crucial role in lowering the freezing point of water, preventing the formation of harmful ice crystals, and preserving the integrity of liposomes during low-temperature storage. The absence or ineffective utilisation of cryoprotectants during extended freezing, particularly in clinical trials, can lead to the destabilisation of liposomes (**Figure 4B**) [111].

Lyophilisation, or freeze-drying, has emerged as a pivotal method in enhancing the stability of liposomal drugs, simplifying their storage and transportation, and extending product shelf life [98, 112, 113]. While research has predominantly focused on non-permeating CPAs such as carbohydrates and sugars in liposome lyophilisation, there is limited information regarding using permeating CPAs other than DMSO in this process. Studies have consistently shown that increasing the concentration of cryoprotectants improves liposomal stability, thereby reducing aggregation [114–117]. These findings underscore the importance of cryoprotectants in enhancing the stability and preservation of liposomal formulations [118–120].

The stability of liposomes in low pH can also be enhanced through complexation to form liposome complexes. PEG is a commonly used polymer that increases the longevity of liposomes *in vitro* and *in vivo* [4, 121–123]. When further modified, PEG can also enhance the thermal and pH stability of liposomes [124–127]. For example, when PEG is complexed with stearoyl and methacryloyl sulfadimethoxine (SDM), the transformation from hydrophilic to a hydrophobic property occurs at pH 7.0. The stearoyl-PEG-poly(SDM) polymers dissolve above pH 7.0 and form aggregates below pH 7.0 [128]. By ionising SDM at pH 6.5 (pH of tumour), the surface of the liposome will become hydrophobic and start aggregating on the surface of the tumour cells. This prevents liposomes from moving back into the bloodstream, increasing their effect on the tumour cells. Further modifications, such as adding thiol and maleimide groups to PEG, create a targeted delivery system through photo-triggered complexation, increasing cellular uptake [129].

5. Conclusion and recommendations

Liposomes are promising drug delivery systems whose potential use and versatility are unequalled in drug delivery. To date, liposomes have proven to be efficient drug delivery systems, with a notable number of formulations being clinically translated. Despite their promising role, major limitations associated with their use have been highlighted. These concerns include but are not limited to poor stability, high biodistribution, and lack of site specificity. Circumventing the limitations associated with liposome surface modification has proven to be a promising strategy that may significantly alter the use of conventional liposomes. Currently, surface modification strategies include traditional conjugation, which incorporates chemistry protocols and the use of diverse ligands to improve the functionality of liposomes. Surface modification using the conjugation chemistry technique has efficiently improved the controlled release and targeted delivery of drugs and has set a promising foundation for theragnostics. In addition, the use of ligands and small molecules has shown to be potentially effective in the precise delivery of drugs and the functionalisation of liposomes. Although commendable advances have been made regarding the surface modification of liposomes, more studies are still needed to explore and substantiate the surface modification of liposomes thoroughly and to necessitate their translation into clinical settings. More *in vivo*, especially bioimaging studies, are needed to depict the safety and efficacy of the surface modification of liposomes.

Acknowledgements

We would like to thank Gunda Spingies for proofreading and editing the manuscript.

Funding

This research was not funded with an external research grant.

Conflict of interest

The authors declare no conflict of interest.

Dedication

In memory of our mentor, colleague, and friend, Professor Rui Krause, who was instrumental in writing this review and who touched all our lives in a very special and profound way. Rest in peace, Rui.

Author details

Justin B. Safari[1,2,3], Paula Maseko[1], Pathy B. Lokole[1,4,5], Galilée G. Byamungu[1,5,6], Simnikiwe Nogqala[1], Tanaka Ndongwe[7], Jonathan M. Mukanya[1,4,5], Frank Ssengooba[7], Gauta Gold Matlou[8], Xavier Siwe-Noundou[7], Christian I. Nkanga[4,5], Brett I. Pletschke[2] and Rui W.M. Krause[1,9]*

1 Faculty of Science, Department of Chemistry, Rhodes University, Makhanda, South Africa

2 Enzyme Science Programme (ESP), Department of Biochemistry, Microbiology and Bioinformatics, Rhodes University, Makhanda, South Africa

3 Faculty of Pharmaceutical Sciences and Public Health, Department of Pharmacy, Official University of Bukavu, Bukavu, Democratic Republic of the Congo

4 Faculty of Pharmaceutical Sciences, Department of Medicinal Chemistry and Pharmacognosy, University of Kinshasa, Kinshasa, Democratic Republic of the Congo

5 Centre de Recherche en Nanotechnologies Appliquées aux Produits Naturels (CReNAPN), Faculty of Pharmaceutical Sciences, Department of Medicinal Chemistry and Pharmacognosy, University of Kinshasa, Kinshasa, Democratic Republic of the Congo

6 Faculty of Sciences, Department of Chemistry, University of Kinshasa, Kinshasa, Democratic Republic of the Congo

7 Department of Pharmaceutical Sciences, Sefako Makgatho Health Sciences University, Pretoria, South Africa

8 Electron Microscopy Unit, Sefako Makgatho Health Sciences University, Pretoria, South Africa

9 Centre for Chemico- and Biomedicinal Research (CCBR), Faculty of Science, Rhodes University, Makhanda, South Africa

*Address all correspondence to: r.krause@ru.ac.za

References

[1] Jacob S, Nair AB, Shah J, Gupta S, Boddu SHS, Sreeharsha N, et al. Lipid nanoparticles as a promising drug delivery carrier for topical ocular therapy; An overview on recent advances. Pharmaceutics. 2022;**14**:533. DOI: 10.3390/pharmaceutics14030533

[2] Lee J, Yang JH, Kim KS, Park GD. Effect of liposome storage of cyanocobalamin on its degradation by ascorbic acid. Food Supplements and Biomaterials for Health. 2021;**1**:e7. DOI: 10.52361/fsbh.2021.1.e7

[3] Kim HT, Lee J, Jo YJ, Choi MJ. Application of liposome encapsulating lactobacillus curvatus extract in cosmetic emulsion lotion. Materials (Basel). 2021;**14**:7571. DOI: 10.3390/ma14247571

[4] Nag OK, Awasthi V. Surface engineering of liposomes for stealth behavior. Pharmaceutics. 2013;**5**:542-569. DOI: 10.3390/PHARMACEUTICS5040542

[5] Safari JB, Mona LB, Sekaleli BT, Avudi BKN, Isamura BK, Mukubwa GK, et al. Inclusion complexation and liposomal encapsulation of an isoniazid hydrazone derivative in cyclodextrin for PH-dependent controlled release. Journal of Drug Delivery Science and Technology. 2023;**81**:1-12. DOI: 10.1016/j.jddst.2023.104302

[6] Nguyen TX, Huang L, Liu L, Elamin Abdalla AM, Gauthier M, Yang G. Chitosan-coated nano-liposomes for the oral delivery of berberine hydrochloride. Journal of Materials Chemistry B. 2014;*2*:7149-7159. DOI: 10.1039/c4tb00876f

[7] Hua S, Wu SY. The use of lipid-based nanocarriers for targeted pain therapies. Frontiers in Pharmacology. 2013;**4**:69399. DOI: 10.3389/FPHAR.2013.00143/BIBTEX

[8] Ding BS, Dziubla T, Shuvaev VV, Muro S, Muzykantov VR. Advanced drug delivery systems that target the vascular endothelium. Molecular Interventions. 2006;**6**:98-112

[9] Kommineni N, Chaudhari R, Conde J, Tamburaci S, Cecen B, Chandra P, et al. Engineered liposomes in interventional theranostics of solid tumors. ACS Biomaterials Science and Engineering. 2023;**9**:4527-4557. DOI: 10.1021/acsbiomaterials.3c00510

[10] Mougenot MF, Pereira VS, Costa ALR, Lancellotti M, Porcionatto MA, da Silveira JC, et al. Biomimetic nanovesicles—Sources, design, production methods, and applications. Pharmaceutics. 2022;**14**:1-25. DOI: 10.3390/pharmaceutics14102008

[11] Farooq MA, Trevaskis NL. TPGS decorated liposomes as multifunctional nano-delivery systems. Pharmaceutical Research. 2023;**40**:245-263. DOI: 10.1007/s11095-022-03424-6

[12] de Lima PHC, Butera AP, Cabeça LF, Ribeiro-Viana RM. Liposome surface modification by phospholipid chemical reactions. Chemistry and Physics of Lipids. 2021;**237**:105084. DOI: 10.1016/j.chemphyslip.2021.105084

[13] Zhang T, Xu X, Pan Y, Yang H, Han J, Liu J, et al. Specific surface modification of liposomes for gut targeting of food bioactive agents. Comprehensive Reviews in Food Science and Food Safety. 2023;**22**:3685-3706. DOI: 10.1111/1541-4337.13224

[14] Sriwidodo, Umar AK, Wathoni N, Zothantluanga JH, Das S, Luckanagul JA. Liposome-polymer complex for drug delivery system and vaccine stabilization. Heliyon. 2022;**8**:e08934. DOI: 10.1016/j.heliyon.2022.e08934

[15] Yang L, Lin X, Zhou J, Hou S, Fang Y, Bi X, et al. Cell membrane-biomimetic coating via click-mediated liposome fusion for mitigating the foreign-body reaction. Biomaterials. 2021;**271**:120768. DOI: 10.1016/j.biomaterials.2021.120768

[16] Hermanson GT. Preparation of Liposome Conjugates and Derivatives. San Diego, CA: Academic Press; 1996. pp. 858-899

[17] Chavda VP, Balar PC, Teli D, Davidson M, Bojarska J, Apostolopoulos V. Antibody–biopolymer conjugates in oncology: A review. Molecules. 2023;**28**:1-9. DOI: 10.3390/molecules28062605

[18] Eras A, Castillo D, Suárez M, Vispo NS, Albericio F, Rodriguez H. Chemical conjugation in drug delivery systems. Frontiers in Chemistry. 2022;**10**:1-23. DOI: 10.3389/fchem.2022.889083

[19] Eroğlu İ, İbrahim M. Liposome–ligand conjugates: A review on the current state of art. Journal of Drug Targeting. 2020;**28**:225-244. DOI: 10.1080/1061186X.2019.1648479

[20] Cho E, Mun SJ, Jeon M, Kim HK, Baek H, Ham YS, et al. Tumor-targeted liposomes with platycodin D2 promote apoptosis in colorectal cancer. Materials Today Bio. 2023;**22**:100745. DOI: 10.1016/j.mtbio.2023.100745

[21] Grüttner C, Müller K, Teller J, Westphal F, Foreman A, Ivkov R. Synthesis and antibody conjugation of magnetic nanoparticles with improved specific power absorption rates for alternating magnetic field cancer therapy. Journal of Magnetism and Magnetic Materials. 2007;**311**:181-186. DOI: 10.1016/j.jmmm.2006.10.1151

[22] Buckley C, Murphy EJ, Montgomery TR, Major I. Hyaluronic acid: A review of the drug delivery capabilities of this naturally occurring polysaccharide. Polymers (Basel). 2022;**14**:3442. DOI: 10.3390/polym14173442

[23] Daraee H, Etemadi A, Kouhi M, Alimirzalu S, Akbarzadeh A. Application of liposomes in medicine and drug delivery. Artificial Cells, Nanomedicine, and Biotechnology. 2016;**44**:381-391. DOI: 10.3109/21691401.2014.953633

[24] Riaz MK, Riaz MA, Zhang X, Lin C, Wong KH, Chen X, et al. Surface functionalization and targeting strategies of liposomes in solid tumor therapy: A review. International Journal of Molecular Sciences. 2018;**19**:195. DOI: 10.3390/ijms19010195

[25] Hueppe N, Wurm FR, Landfester K. Nanocarriers with multiple cargo load—A comprehensive preparation guideline using orthogonal strategies. Macromolecular Rapid Communications. 2023;**44**:1-18. DOI: 10.1002/marc.202200611

[26] Alazawi S, Al-Bayati SM, Rasheed AM. Click chemistry complexes as a platform for biological application: A review. Advances in Natural Sciences: Nanoscience and Nanotechnology. 2022;**1**:121-140. DOI: 10.55945/joasnt.2022.1.4.121-140

[27] Chen Z, Wang Z, Gu Z. Bioinspired and biomimetic nanomedicines. Accounts of Chemical Research. 2019;**52**:1255-1264. DOI: 10.1021/acs.accounts.9b00079

[28] Rahamim V, Azagury A. Bioengineered biomimetic and bioinspired noninvasive drug delivery systems. Advanced Functional Materials. 2021;**31**:1-17. DOI: 10.1002/adfm.202102033

[29] Balmert SC, Little SR. Biomimetic delivery with micro- and nanoparticles. Advanced Materials. 2012;**24**:3757-3778. DOI: 10.1002/adma.201200224

[30] Meyer RA, Sunshine JC, Green JJ. Biomimetic particles as therapeutics. Trends in Biotechnology. 2015;**33**:514-524. DOI: 10.1016/j.tibtech.2015.07.001

[31] Barba-Bon A, Nilam M, Hennig A. Supramolecular chemistry in the biomembrane. Chembiochem. 2020;**21**:886-910. DOI: 10.1002/cbic.201900646

[32] Giuliano CB, Cvjetan N, Ayache J, Walde P. Multivesicular vesicles: Preparation and applications. ChemSystemsChem. 2021;**3**:e2000049. DOI: 10.1002/syst.202000049

[33] Yang J, Bahreman A, Daudey G, Bussmann J, Olsthoorn RCL, Kros A. Drug delivery via cell membrane fusion using lipopeptide modified liposomes. ACS Central Science. 2016;**2**:621-630. DOI: 10.1021/acscentsci.6b00172

[34] Alies B, Ouelhazi MA, Noireau A, Gaudin K, Barthélémy P. Silver ions detection via nucleolipids self-assembly. Analytical Chemistry. 2019;**91**:1692-1695. DOI: 10.1021/acs.analchem.8b04066

[35] Chen S, Chen R. A virus-mimicking, endosomolytic liposomal system for efficient, PH-triggered intracellular drug delivery. ACS Applied Materials and Interfaces. 2016;**8**:22457-22467. DOI: 10.1021/acsami.6b05041

[36] Cavalcanti RRM, Lira RB, Riske KA. Membrane fusion biophysical analysis of fusogenic liposomes. Langmuir. 2022;**38**:10430-10441. DOI: 10.1021/acs.langmuir.2c01169

[37] Zeng Y, Shen M, Singhal A, Sevink GJA, Crone N, Boyle AL, et al. Enhanced liposomal drug delivery via membrane fusion triggered by dimeric coiled-coil peptides. Small. 2023;**19**:1-14. DOI: 10.1002/smll.202301133

[38] Lira RB, Robinson T, Dimova R, Riske KA. Highly efficient protein-free membrane fusion: A giant vesicle study. Biophysical Journal. 2019;**116**:79-91. DOI: 10.1016/j.bpj.2018.11.3128

[39] Pignatello R, Nicolosi D, Nicolosi VM. Fusogenic liposomes as new carriers to enlarge the spectrum of action of antibiotic drugs against gram-negative bacteria. Science Against Microbial Pathogens: Communicating Current Research and Technological Advances. 2011;**1**:52-60

[40] Kakudo T, Chaki S, Futaki S, Nakase I, Akaji K, Kawakami T, et al. Transferrin-modified liposomes equipped with a PH-sensitive fusogenic peptide: An artificial viral-like delivery system. Biochemistry. 2004;**43**:5618-5628. DOI: 10.1021/bi035802w

[41] Yang ST, Zaitseva E, Chernomordik LV, Melikov K. Cell-penetrating peptide induces leaky fusion of liposomes containing late endosome-specific anionic lipid. Biophysical Journal. 2010;**99**:2525-2533. DOI: 10.1016/j.bpj.2010.08.029

[42] Pak CC, Ali S, Janoff AS, Meers P. Triggerable liposomal fusion by enzyme cleavage of a novel peptide-lipid conjugate. Biochimica et Biophysica Acta (BBA)—Biomembranes. 1998;**1372**:13-27. DOI: 10.1016/S0005-2736(98)00041-8

[43] Motion JPM, Nguyen J, Szoka FC. Phosphatase-triggered

fusogenic liposomes for cytoplasmic delivery of cell-impermeable compounds. Angewandte Chemie. 2012;**124**:9181-9185. DOI: 10.1002/ange.201204198

[44] Bird RE, Lemmel SA, Yu X, Zhou QA. Bioorthogonal chemistry and its applications. Bioconjugate Chemistry. 2021;**32**:2457-2479. DOI: 10.1021/acs.bioconjchem.1c00461

[45] Devaraj NK. The future of bioorthogonal chemistry. ACS Central Science. 2018;**4**:952-959. DOI: 10.1021/acscentsci.8b00251

[46] Elahipanah S, Radmanesh P, Luo W, O'Brien PJ, Rogozhnikov D, Yousaf MN. Rewiring gram-negative bacteria cell surfaces with bio-orthogonal chemistry via liposome fusion. Bioconjugate Chemistry. 2016;**27**:1082-1089. DOI: 10.1021/acs.bioconjchem.6b00073

[47] Emmetiere F, Irwin C, Viola-Villegas NT, Longo V, Cheal SM, Zanzonico P, et al. 18F-labeled-bioorthogonal liposomes for in vivo targeting. Bioconjugate Chemistry. 2013;**24**:1784-1789. DOI: 10.1021/bc400322h

[48] Gai M, Simon J, Lieberwirth I, Mailänder V, Morsbach S, Landfester K. A bio-orthogonal functionalization strategy for site-specific coupling of antibodies on vesicle surfaces after self-assembly. Polymer Chemistry. 2020;**11**:527-540. DOI: 10.1039/c9py01136f

[49] Nakamura K, Yamashita K, Itoh Y, Yoshino K, Nozawa S, Kasukawa H. Comparative studies of polyethylene glycol-modified liposomes prepared using different PEG-modification methods. Biochimica et Biophysica Acta (BBA)—Biomembranes. 1818;**2012**:2801-2807. DOI: 10.1016/j.bbamem.2012.06.019

[50] Feldborg LN, Jølck RI, Andresen TL. Quantitative evaluation of bioorthogonal chemistries for surface functionalization of nanoparticles. Bioconjugate Chemistry. 2012;**23**:2444-2450. DOI: 10.1021/bc3005057

[51] Rideau E, Dimova R, Schwille P, Wurm FR, Landfester K. Liposomes and polymersomes: A comparative review towards cell mimicking. Chemical Society Reviews. 2018;**47**:8572-8610. DOI: 10.1039/c8cs00162f

[52] Li X, Rao X, Cai L, Liu X, Wang H, Wu W, et al. Targeting tumor cells by natural anti-carbohydrate antibodies using rhamnose-functionalized liposomes. ACS Chemical Biology. 2016;**11**:1205-1209. DOI: 10.1021/acschembio.6b00173

[53] Gong YK, Winnik FM. Strategies in biomimetic surface engineering of nanoparticles for biomedical applications. Nanoscale. 2012;**4**:360-368. DOI: 10.1039/c1nr11297j

[54] El Maghraby GM, Barry BW, Williams AC. Liposomes and skin: From drug delivery to model membranes. European Journal of Pharmaceutical Sciences. 2008;**34**:203-222. DOI: 10.1016/j.ejps.2008.05.002

[55] Chen H, Zheng Y, Jiang JH, Wu HL, Shen GL, Yu RQ. An ultrasensitive chemiluminescence biosensor for cholera toxin based on ganglioside-functionalized supported lipid membrane and liposome. Biosensors and Bioelectronics. 2008;**24**:684-689. DOI: 10.1016/j.bios.2008.06.031

[56] Weingart OG, Loessner MJ. Nerve cell-mimicking liposomes as biosensor for botulinum neurotoxin complete physiological activity. Toxicology and Applied Pharmacology. 2016;**313**:16-23. DOI: 10.1016/j.taap.2016.10.010

[57] Wood JC, Fassler JD, Meade T. Mimicking liver iron overload using liposomal ferritin preparations. Magnetic Resonance in Medicine. 2004;**51**:607-611. DOI: 10.1002/mrm.10735

[58] Unger WWJ, Van Beelen AJ, Bruijns SC, Joshi M, Fehres CM, Van Bloois L, et al. Glycan-modified liposomes boost CD4 + and CD8 + T-cell responses by targeting DC-SIGN on dendritic cells. Journal of Controlled Release. 2012;**160**:88-95. DOI: 10.1016/j.jconrel.2012.02.007

[59] Lepenies B, Lee J, Sonkaria S. Targeting C-type lectin receptors with multivalent carbohydrate ligands. Advanced Drug Delivery Reviews. 2013;**65**:1271-1281. DOI: 10.1016/j.addr.2013.05.007

[60] Tang B, Qian Y, Gou Y, Cheng G, Fang G. VE-albumin core-shell nanoparticles for paclitaxel delivery to treat MDR breast cancer. Molecules. 2018;**23**:2760. DOI: 10.3390/molecules23112760

[61] Shehata T, Ogawara KI, Higaki K, Kimura T. Prolongation of residence time of liposome by surface-modification with mixture of hydrophilic polymers. International Journal of Pharmaceutics. 2008;**359**:272-279. DOI: 10.1016/j.ijpharm.2008.04.004

[62] Miao YQ, Chen MS, Zhou X, Guo LM, Zhu JJ, Wang R, et al. Chitosan oligosaccharide modified liposomes enhance lung cancer delivery of paclitaxel. Acta Pharmacologica Sinica. 2021;**42**:1714-1722. DOI: 10.1038/s41401-020-00594-0

[63] Kono Y, Iwasaki A, Fujita T. Effect of surface charge, particle size, and modification by polyethylene glycol of liposomes on their association with caco-2 cells across an unstirred water layer. Pharmazie. 2018;**73**:3-8. DOI: 10.1691/ph.2018.7110

[64] Hama S, Sakai M, Itakura S, Majima E, Kogure K. Rapid modification of antibodies on the surface of liposomes composed of high-affinity protein A-conjugated phospholipid for selective drug delivery. Biochemistry and Biophysics Reports. 2021;**27**:101067. DOI: 10.1016/j.bbrep.2021.101067

[65] Ye Z, Gastfriend BD, Umlauf BJ, Lynn DM, Shusta EV. Antibody-targeted liposomes for enhanced targeting of the blood-brain barrier. Pharmaceutical Research. 2022;**39**:1523-1534. DOI: 10.1007/s11095-022-03186-1

[66] Drabovich AP, Berezovski MV, Musheev MU, Krylov SN. Selection of smart small-molecule ligands: The proof of principle. Analytical Chemistry. 2009;**81**:490-494. DOI: 10.1021/ac8023813

[67] Nobs L, Buchegger F, Gurny R, Allémann E. Current methods for attaching targeting ligands to liposomes and nanoparticles. Journal of Pharmaceutical Sciences. 2004;**93**:1980-1992. DOI: 10.1002/jps.20098

[68] Jain SK, Jain A. Ligand mediated drug targeted liposomes. Color Research and Application. 2002;**27**:379. DOI: 10.1002/col.10105

[69] Khan AA, Allemailem KS, AlmatroodiSA, AlmatroudiA, RahmaniAH. Recent strategies towards the surface modification of liposomes: An innovative approach for different clinical applications. 3 Biotech. 2020;**10**:1-15. DOI: 10.1007/s13205-020-2144-3

[70] Sriram D, Yogeeswari P, Madhu K. Synthesis and in vitro and in vivo antimycobacterial activity of isonicotinoyl hydrazones. Bioorganic

and Medicinal Chemistry Letters. 2005;**15**:4502-4505. DOI: 10.1016/j.bmcl.2005.07.011

[71] Gabizon A, Horowitz AT, Goren D, Tzemach D, Mandelbaum-Shavit F, Qazen MM, et al. Targeting folate receptor with folate linked to extremities of poly(ethylene glycol)-grafted liposomes: In vitro studies. Bioconjugate Chemistry. 1999;**10**:289-298. DOI: 10.1021/bc9801124

[72] Aronson MR, Medina SH, Mitchell MJ. Peptide functionalized liposomes for receptor targeted cancer therapy. APL Bioengineering. 2021;**5**:011501. DOI: 10.1063/5.0029860

[73] Belfiore L, Saunders DN, Ranson M, Thurecht KJ, Storm G, Vine KL. Towards clinical translation of ligand-functionalized liposomes in targeted cancer therapy: Challenges and opportunities. Journal of Controlled Release. 2018;**277**:1-13. DOI: 10.1016/j.jconrel.2018.02.040

[74] Moulahoum H, Ghorbanizamani F, Zihnioglu F, Timur S. Surface biomodification of liposomes and polymersomes for efficient targeted drug delivery. Bioconjugate Chemistry. 2021;**32**:1491-1502. DOI: 10.1021/acs.bioconjchem.1c00285

[75] Urbiola K, Blanco-Fernández L, Ogris M, Rödl W, Wagner E, de Ilarduya CT. Novel PAMAM-PEG-peptide conjugates for SiRNA delivery targeted to the transferrin and epidermal growth factor receptors. Journal of Personalized Medicine. 2018;**8**:4. DOI: 10.3390/jpm8010004

[76] Carvalheiro M, Ferreira-Silva M, Holovanchuk D, Marinho HS, Moreira JN, Soares H, et al. Antagonist G-targeted liposomes for improved delivery of anticancer drugs in small cell lung carcinoma. International Journal of Pharmaceutics. 2022;**612**:121380. DOI: 10.1016/j.ijpharm.2021.121380

[77] Di J, Xie F, Xu Y. When liposomes met antibodies: Drug delivery and beyond. Advanced Drug Delivery Reviews. 2020;**154-155**:151-162. DOI: 10.1016/j.addr.2020.09.003

[78] Richards DA, Maruani A, Chudasama V. Antibody fragments as nanoparticle targeting ligands: A step in the right direction. Chemical Science. 2016;**8**:63-77. DOI: 10.1039/c6sc02403c

[79] Kedmi R, Veiga N, Ramishetti S, Goldsmith M, Rosenblum D, Dammes N, et al. A modular platform for targeted RNAi therapeutics. Nature Nanotechnology. 2018;**13**:214-219. DOI: 10.1038/s41565-017-0043-5

[80] Reulen SWA, Brusselaars WWT, Langereis S, Mulder WJM, Breurken M, Merkx M. Protein-liposome conjugates using cysteine-lipids and native chemical ligation. Bioconjugate Chemistry. 2007;**18**:590-596. DOI: 10.1021/bc0602782

[81] Fleiner M, Benzinger P, Fichert T, Massing U. Studies on protein-liposome coupling using novel thiol-reactive coupling lipids: Influence of spacer length and polarity. Bioconjugate Chemistry. 2001;**12**:470-475. DOI: 10.1021/bc000101m

[82] Nakano K, Tozuka Y, Takeuchi H. Effect of surface properties of liposomes coated with a modified polyvinyl alcohol (PVA-R) on the interaction with macrophage cells. International Journal of Pharmaceutics. 2008;**354**:174-179. DOI: 10.1016/j.ijpharm.2007.08.046

[83] Le NTT, Du Cao V, Nguyen TNQ, Le TTH, Tran TT, Thi TTH. Soy lecithin-derived liposomal delivery systems:

Surface modification and current applications. International Journal of Molecular Sciences. 2019;**20**:4706. DOI: 10.3390/ijms20194706

[84] Nkanga CI, Krause RWM. Encapsulation of isoniazid-conjugated phthalocyanine-in-cyclodextrin-in-liposomes using heating method. Scientific Reports. 2019;**9**:1-16. DOI: 10.1038/s41598-019-47991-y

[85] Shimizu T, Abu Lila AS, Fujita R, Awata M, Kawanishi M, Hashimoto Y, et al. A hydroxyl PEG version of PEGylated liposomes and its impact on anti-PEG IgM induction and on the accelerated clearance of PEGylated liposomes. European Journal of Pharmaceutics and Biopharmaceutics. 2018;**127**:142-149. DOI: 10.1016/j.ejpb.2018.02.019

[86] Hollmann A, Delfederico L, Glikmann G, De Antoni G, Semorile L, Disalvo EA. Characterization of liposomes coated with S-layer proteins from lactobacilli. Biochimica et Biophysica Acta (BBA)—Biomembranes. 2007;**1768**:393-400. DOI: 10.1016/j.bbamem.2006.09.009

[87] Tran TTT, Tran VH, Lam TT. Encapsulation of tagitinin C in liposomes coated by tithonia diversifolia pectin. Journal of Microencapsulation. 2019;**36**:53-61. DOI: 10.1080/02652048.2019.1585983

[88] Double JA, Bibby MC. Therapeutic index: A vital component. Journal of the National Cancer Institute. 1989;**81**:988-994

[89] Peng P, Chen Z, Wang M, Wen B, Deng X. Polysaccharide-modified liposomes and their application in cancer research. Chemical Biology and Drug Design. 2023;**101**:998-1011. DOI: 10.1111/cbdd.14201

[90] Cano ME, Lesur D, Bincoletto V, Gazzano E, Stella B, Riganti C, et al. Synthesis of defined oligohyaluronates-decorated liposomes and interaction with lung cancer cells. Carbohydrate Polymers. 2020;**248**:116798. DOI: 10.1016/j.carbpol.2020.116798

[91] Ding Y, Yang R, Yu W, Hu C, Zhang Z, Liu D, et al. Chitosan oligosaccharide decorated liposomes combined with TH302 for photodynamic therapy in triple negative breast cancer. Journal of Nanobiotechnology. 2021;**19**:1-17. DOI: 10.1186/s12951-021-00891-8

[92] Alavi S, Haeri A, Dadashzadeh S. Utilization of chitosan-caged liposomes to push the boundaries of therapeutic delivery. Carbohydrate Polymers. 2017;**157**:991-1012

[93] Kaur CD, Nahar M, Jain NK. Lymphatic targeting of zidovudine using surface-engineered liposomes. Journal of Drug Targeting. 2008;**16**:798-805. DOI: 10.1080/10611860802475688

[94] Marie H, Lemaire L, Franconi F, Lajnef S, Frapart YM, Nicolas V, et al. Superparamagnetic liposomes for MRI monitoring and external magnetic field-induced selective targeting of malignant brain tumors. Advanced Functional Materials. 2015;**25**:1258-1269. DOI: 10.1002/adfm.201402289

[95] Riaz MK, Zhang X, Wong KH, Chen H, Liu Q, Chen X, et al. Pulmonary delivery of transferrin receptors targeting peptide surface-functionalized liposomes augments the chemotherapeutic effect of quercetin in lung cancer therapy. International Journal of Nanomedicine. 2019;**14**:2879-2902. DOI: 10.2147/IJN.S192219

[96] Shahin M, Soudy R, Aliabadi HM, Kneteman N, Kaur K, Lavasanifar A. Engineered breast tumor targeting

peptide ligand modified liposomal doxorubicin and the effect of peptide density on anticancer activity. Biomaterials. 2013;**34**:4089-4097. DOI: 10.1016/j.biomaterials.2013.02.019

[97] Weber C, Voigt M, Simon J, Danner AK, Frey H, Mailänder V, et al. Functionalization of liposomes with hydrophilic polymers results in macrophage uptake independent of the protein corona. Biomacromolecules. 2019;**20**:2989-2999. DOI: 10.1021/acs.biomac.9b00539

[98] Wang Y, Grainger DW. Lyophilized liposome-based parenteral drug development: Reviewing complex product design strategies and current regulatory environments. Advanced Drug Delivery Reviews. 2019;**151-152**:56-71

[99] Liu W, Ye A, Han F, Han J. Advances and challenges in liposome digestion: Surface interaction, biological fate, and GIT modeling. Advances in Colloid and Interface Science. 2019;**263**:52-67

[100] Kapoor M, Lee SL, Tyner KM. Liposomal drug product development and quality: Current US experience and perspective. The AAPS Journal. 2017;**19**:632-641. DOI: 10.1208/s12248-017-0049-9

[101] Inglut CT, Sorrin AJ, Kuruppu T, Vig S, Cicalo J, Ahmad H, et al. Immunological and toxicological considerations for the design of liposomes. Nanomaterials. 2020;**10**:190. DOI: 10.3390/nano10020190

[102] Lu RM, Chen MS, Chang DK, Chiu CY, Lin WC, Yan SL, et al. Targeted drug delivery systems mediated by a novel peptide in breast cancer therapy and imaging. PLoS One. 2013;**8**:e66128. DOI: 10.1371/journal.pone.0066128

[103] Yingchoncharoen P, Kalinowski DS, Richardson DR. Lipid-based drug delivery systems in cancer therapy: What is available and what is yet to come. Pharmacological Reviews. 2016;**68**:701-787. DOI: 10.1124/pr.115.012070

[104] Maranhão RC, Vital CG, Tavoni TM, Graziani SR. Clinical experience with drug delivery systems as tools to decrease the toxicity of anticancer chemotherapeutic agents. Expert Opinion on Drug Delivery. 2017;**14**:1217-1226

[105] Lee Y, Thompson DH. Stimuli-responsive liposomes for drug delivery. Wiley Interdisciplinary. Reviews Nanomedicine Nanobiotechnology. 2017;**9**:e1450. DOI: 10.1002/wnan.1450

[106] Lee DE, Lew MG, Woodbury DJ. Vesicle fusion to planar membranes is enhanced by cholesterol and low temperature. Chemistry and Physics of Lipids. 2013;**166**:45-54. DOI: 10.1016/j.chemphyslip.2012.11.004

[107] Gibis M, Zeeb B, Weiss J. Formation, characterization, and stability of encapsulated hibiscus extract in multilayered liposomes. Food Hydrocolloids. 2014;**38**:28-39. DOI: 10.1016/j.foodhyd.2013.11.014

[108] Sepúlveda CT, Alemán A, Zapata JE, Montero MP, Gómez-Guillén MC. Characterization and storage stability of spray dried soy-rapeseed lecithin/trehalose liposomes loaded with a tilapia viscera hydrolysate. Innovative Food Science and Emerging Technologies. 2021;**71**:102708. DOI: 10.1016/j.ifset.2021.102708

[109] Subramanian P, Kurek MA, Jensen I-J. Lipid-based nanocarrier system for the effective delivery of nutraceuticals. Molecules. 2021;**26**:5510. DOI: 10.3390/MOLECULES26185510

[110] He H, Lu Y, Qi J, Zhu Q, Chen Z, Wu W. Adapting liposomes for oral drug delivery. Acta Pharmaceutica Sinica B. 2019;**9**:36-48. DOI: 10.1016/j.apsb.2018.06.005

[111] Gregoriadis G. Liposome Technology: Volume III: Targeted Drug Delivery and Biological Interaction. Boca Raton: CRC Press; 2018. ISBN 9781351082570

[112] Stark B, Pabst G, Prassl R. Long-term stability of sterically stabilized liposomes by freezing and freeze-drying: Effects of cryoprotectants on structure. European Journal of Pharmaceutical Sciences. 2010;**41**:546-555. DOI: 10.1016/J.EJPS.2010.08.010

[113] Franzé S, Selmin F, Samaritani E, Minghetti P, Cilurzo F. Lyophilization of liposomal formulations: Still necessary, still challenging. Pharmaceutics. 2018;**10**:139. DOI: 10.3390/PHARMACEUTICS10030139

[114] Suzuki T, Komatsu H, Miyajima K. Effects of glucose and its oligomers on the stability of freeze-dried liposomes. Biochimica et Biophysica Acta (BBA)—Biomembranes. 1996;**1278**:176-182. DOI: 10.1016/0005-2736(95)00221-9

[115] Morandi MI, Sommer M, Kluzek M, Thalmann F, Schroder AP, Marques CM. DPPC bilayers in solutions of high sucrose content. Biophysical Journal. 2018;**114**:2165-2173. DOI: 10.1016/j.bpj.2018.04.003

[116] Hinrichs WLJ, Sanders NN, De Smedt SC, Demeester J, Frijlink HW. Inulin is a promising cryo- and lyoprotectant for PEGylated lipoplexes. Journal of Controlled Release. 2005;**103**:465-479. DOI: 10.1016/J.JCONREL.2004.12.011

[117] Sydykov B, Oldenhof H, de Oliveira Barros L, Sieme H, Wolkers WF. Membrane permeabilization of phosphatidylcholine liposomes induced by cryopreservation and vitrification solutions. Biochimica et Biophysica Acta (BBA)—Biomembranes. 1860;**2018**:467-474. DOI: 10.1016/J.BBAMEM.2017.10.031

[118] Smits AJ, Albrecht B, Ag LK. Glycerol-based liposomal systems. Natural Ingredients. 2014;**2014**:11

[119] Koster KL, Webb MS, Bryant G, Lynch DV. Interactions between soluble sugars and POPC (1-palmitoyl-2-oleoylphosphatidylcholine) during dehydration: Vitrification of sugars alters the phase behavior of the phospholipid. Biochimica et Biophysica Acta (BBA)—Biomembranes. 1994;**1193**:143-150. DOI: 10.1016/0005-2736(94)90343-3

[120] Susa F, Bucca G, Limongi T, Cauda V, Pisano R. Enhancing the preservation of liposomes: The role of cryoprotectants, lipid formulations and freezing approaches. Cryobiology. 2021;**98**:46-56. DOI: 10.1016/J.CRYOBIOL.2020.12.009

[121] Torchilin VP. Multifunctional nanocarriers. Advanced Drug Delivery Reviews. 2006;**58**:1532-1555. DOI: 10.1016/J.ADDR.2006.09.009

[122] Petros RA, Desimone JM. Strategies in the design of nanoparticles for therapeutic applications. Nature Reviews. Drug Discovery. 2010;**9**:615-627. DOI: 10.1038/nrd2591

[123] Duggan ST, Keating GM, Ferrandina G, Kesterson JP, Lorusso D, Muggia F. Pegylated liposomal doxorubicin: A review of its use in metastatic breast cancer, ovarian cancer, multiple myeloma and AIDS-related kaposis sarcoma. Drugs. 2011;**71**:2531-2558. DOI: 10.2165/11207510-000000000-00000/FIGURES/TAB7

[124] Chiang YT, Lyu SY, Wen YH, Lo CL. Preparation and characterization of electrostatically crosslinked polymer–liposomes in anticancer therapy. International Journal of Molecular Sciences. 2018;**19**:1615. DOI: 10.3390/ijms19061615

[125] Simões MG, Alves P, Carvalheiro M, Simões PN. Stability effect of cholesterol-poly(acrylic acid) in a stimuli-responsive polymer-liposome complex obtained from soybean lecithin for controlled drug delivery. Colloids Surfaces B Biointerfaces. 2017;**152**:103-113. DOI: 10.1016/j.colsurfb.2017.01.002

[126] Chiang YT, Lo CL. PH-responsive polymer-liposomes for intracellular drug delivery and tumor extracellular matrix switched-on targeted cancer therapy. Biomaterials. 2014;**35**:5414-5424. DOI: 10.1016/j.biomaterials.2014.03.046

[127] Alves P, Hugo AA, Tymczyszyn EE, Ferreira AF, Fausto R, Pérez PF, et al. Effect of cholesterol-poly(N,N-dimethylaminoethyl methacrylate) on the properties of stimuli-responsive polymer liposome complexes. Colloids Surfaces B Biointerfaces. 2013;**104**:254-261. DOI: 10.1016/j.colsurfb.2012.12.016

[128] Bersani S, Vila-Caballer M, Brazzale C, Barattin M, Salmaso S. PH-sensitive stearoyl-PEG-poly(methacryloyl sulfadimethoxine) decorated liposomes for the delivery of gemcitabine to cancer cells. European Journal of Pharmaceutics and Biopharmaceutics. 2014;**88**:670-682. DOI: 10.1016/j.ejpb.2014.08.005

[129] Liang Y, Kiick KL. Multifunctional lipid-coated polymer nanogels crosslinked by photo-triggered Michael-type addition. Polymer Chemistry. 2014;**5**:1728-1736. DOI: 10.1039/c3py01269g

Chapter 2

Spontaneous Formation of Vesicular Liposomes: Thermodynamics and Bending Energetics

Lars Magnus Bergström

Abstract

By means of combining bending elasticity theory with solution thermodynamics of small systems, we demonstrate that unilamellar vesicular liposomes can be thermodynamically stable with a wide range of average sizes depending on the various bending elasticity constants. The average vesicle size increases with increasing bending rigidity (k_c) and saddle-splay constant ($\bar{k}_c$), and with decreasing spontaneous curvature (H_0). Bilayer aggregates predominate over micelles at lower values of the spontaneous curvature, in the regime of which H_0 favours large vesicles. However, small unilamellar vesicles may be favoured by low values of k_c, rather than high H_0. Mixing two amphiphilic components with different spontaneous curvatures gives rise to an explicit contribution that always brings down k_c. As a result, the amphiphilic component with high positive spontaneous curvature prefers to be located in the outer, positively curved, monolayer, whereas the other compound prefers to be located in the inner, negatively curved, layer. In contrast to unilamellar vesicles, geometrically open disks can only be thermodynamically stable close to the micelle-to-bilayer transition in a dilute solution of non-interacting bilayer aggregates. However, in more concentrated solutions, above the overpacking limit of vesicles, disks may be more favourable than vesicles due to more favourable packing conditions.

Keywords: vesicles, surfactants, amphiphilic lipids, spontaneous self-assembly, bending elasticity

1. Introduction

Colloidal solutions are thermodynamically stable systems in which amphiphilic molecules (surfactants or amphiphilic lipids) self-assemble spontaneously to micelles, microemulsions, or bilayer aggregates. This is in contrast to colloidal dispersions, which are non-equilibrium systems consisting of one phase dispersed in another (i.e. two phases), like emulsions or dispersed vesicular liposomes. The stability of aggregates in colloidal solutions depends on the local curvature of the aggregates together with the entropy of self-assembling amphiphilic molecules. The latter effect tends to reduce the size of the self-assembled aggregates. As a result, the size and shape of a thermodynamically stable aggregate are determined by the curvature properties of the aggregate interface together with the entropy of self-assembly [1]. A positive spontaneous curvature is required for oil-in-water droplets or ordinary surfactant micelles to form in an aqueous solvent. Likewise, water-in-oil droplets or reversed micelles dissolved in a continuous oil phase may form by amphiphilic molecules with negative spontaneous curvature [2].

The simplest form of aggregate structure is a geometrically homogeneous spherical micelle or microemulsion droplet, with a homogeneous curvature all over its interface. Other aggregate shapes include non-spherical micelles, unilamellar vesicles, and bilayer disks. Such aggregates may be considered as composed of geometrical parts with different curvatures and may form under certain conditions related to the detailed geometrical shape. The formation of non-spherical elongated micelles has recently been theoretically treated in the so-called general micelle model by mean of combining thermodynamics of self-assembly with bending elasticity theory [3]. Bilayer aggregates are expected to form in an aqueous solvent by amphiphilic molecules with low spontaneous curvature (or in an oil phase by amphiphilic molecules with high spontaneous curvature). When mixing a micelle-forming surfactant with a bilayer-forming amphiphilic lipid, a reversible transition from micelles to bilayers is usually observed at a certain surfactant/lipid composition in the aggregates [4].

Bilayer aggregates in isotropic solutions are mainly shaped as either open circular disks or geometrically closed unilamellar vesicles. Both aggregate types are geometrically heterogeneous and considered as composed of different geometrical parts. For instance, a vesicle is defined as a (nearly spherical) geometrically closed bilayer consisting of two geometrical parts: a positively curved outer monolayer and a negatively curved inner monolayer (*cf.* **Figure 1**). A vesicular aggregate is sometimes denoted as a *liposome* when at least one of the components constituting the vesicles is

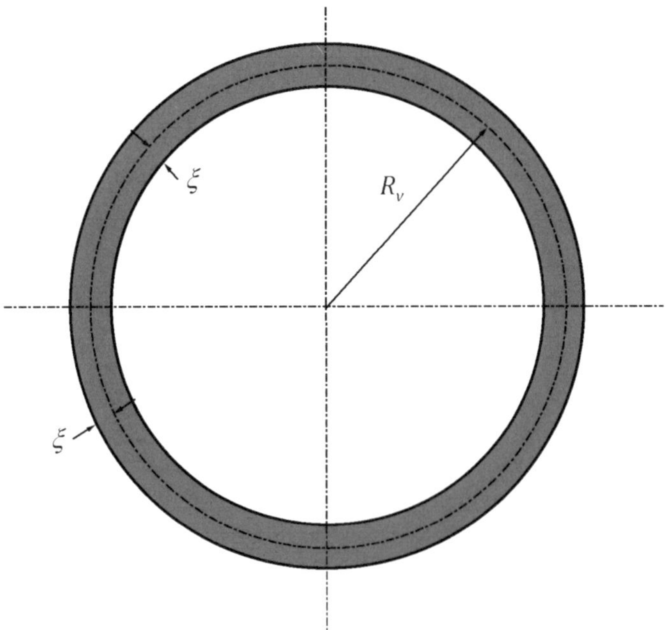

Figure 1.
Schematic picture of a spherical unilamellar vesicles with bilayer midplane radius R_v and monolayer thickness ξ.

an amphiphilic lipid. Liposomes made up of amphiphilic lipids are usually non-equilibrium structures in a colloidal dispersion that need input of energy to be able to form, for example by means of sonication and extrusion. However, by simply mixing two amphiphilic components, unilamellar vesicles sometimes form spontaneously in a thermodynamically stable colloidal solution, depending on the choice of components. Notably, vesicles have been observed to form spontaneously in several systems where two oppositely charged surfactants have been dissolved in water [5]. The size of unilamellar vesicles ranges from about 20 nm in diameter up to several hundreds of nanometer. These mixed cationic/anionic vesicles are usually found in the range 20–100 nm and referred to as *small unilamellar vesicles* (SUV). The size of cationic/anionic vesicles has been found to be sensitive to the chemical structure of components as well as surfactant concentration and ionic strength of an aqueous solvent. These vesicles tend to increase in size with decreasing total surfactant concentration and increasing ionic strength, and they become destabilized and transform into bilayer disks as the aggregates become sufficiently large [6–8].

Spontaneously formed vesicles have also been observed in mixtures of phospholipid and bile salt surfactants [9, 10] and, more recently, in mixtures of phospholipid and certain amphiphilic drug surfactants [11]. In the latter case, ultrasmall vesicles with a diameter as small as less than 20 nm were discovered to form spontaneously. In all these systems, the vesicles were formed spontaneously from mixed micelles by means of simply diluting more concentrated micellar solutions. In contrast to the mixed cationic/anionic surfactant systems, the mixed drug surfactant/phospholipid vesicles are stable at high ionic strengths (physiological saline solution) and decrease, rather than increase, in size upon diluting the samples.

In this chapter, we theoretically investigate the spontaneous formation of unilamellar vesicles by means of combining thermodynamics of self-assembly and bending elasticity theory in a similar manner as has previously been done for micelles in the general micelle model [3, 12] and for microemulsions [2]. In particular, we compare the stability of unilamellar vesicles and bilayer disks from a thermodynamic point of view. We are able to conclude that in sufficiently dilute solutions, where inter-aggregate interactions are neglected, geometrically closed vesicles are the thermodynamically stable aggregate type of bilayers.

2. Thermodynamics of self-assembly

Thermodynamically stable equilibrium structures where N-free amphiphilic molecules (monomers) dissolved in an aqueous phase self-assemble to form a small bilayer system B_N (vesicle or disk) can be treated in terms of a set of multiple equilibrium reactions

$$N \text{ monomers} \rightleftharpoons B_N \tag{1}$$

The change in Gibbs energy in the self-assembly process may be written as a sum of two contributions:

$$\Delta G = N\Delta\mu - T\Delta S_{agg} \tag{2}$$

where the change in entropy for the process in Eq. (1), when self-assembling N free monomers at a volume fraction ϕ_{free} into a bilayer aggregate B_N with volume fraction ϕ_N, is given by

$$\Delta S_{agg} = -k\left(\ln \phi_N - N \ln \phi_{free}\right) \tag{3}$$

and k is Boltzmann's constant. ΔS_{agg} must always be a negative quantity since the process of self-assembly is unfavourable in the absence of a specific driving force for the process in Eq. (1). The quantity $\Delta\mu$ comprises several residual contributions to the free energy of forming a single bilayer aggregate. The most important contribution to $\Delta\mu$ comes from the main driving force for the self-assembly process, the hydrophobic effect; that is, the principle that oil and water do not mix and the hydrocarbon-water interfacial area tends to become reduced.

Combining Eqs. (2) and (3) gives the following set of equilibrium conditions, one for each aggregation number N

$$\Delta G_N = \mathrm{E}_N + kT \ln \phi_N = 0 \tag{4}$$

where T is the absolute temperature. For the sake of simplicity, we have introduced the free energy parameter E_N, defined as [13]

$$\mathrm{E}_N \equiv N\Delta\mu - NkT \ln \phi_{free} \tag{5}$$

At equilibrium, $\Delta G_N = 0$ and amphiphilic molecules are reversibly exchanged between bilayer aggregates and as free monomers. In accordance with Eq. (4), the volume fraction of aggregates with aggregation number N equals

$$\phi_N = e^{-\mathrm{E}_N/kT} \tag{6}$$

Summing up the different volume fractions in Eq. (6) gives the total volume fraction ϕ_{bil} of amphiphilic molecules self-assembled in bilayer aggregates, that is

$$\phi_{bil} = \sum_{N=1}^{\infty} \phi_N \approx \int_1^{\infty} e^{-\mathrm{E}_N/kT} dN \tag{7}$$

Inserting the proper mathematical expression for the function E_N in Eq. (7) gives the size distribution of aggregates. Below we will derive expressions of E_N for bilayer vesicles and disks, respectively, using bending elasticity theory, to arrive at the proper size distribution for vesicular liposomes and bilayer disks.

3. Bending elasticity theory

The free energy per unit area γ at a single point of a self-assembled monolayer depends on the mean and Gaussian curvatures, $H = (c_1 + c_2)/2$ and $K = c_1c_2$, respectively, for a given small system of self-assembled amphiphilic molecules in a particular solvent at a given set of environmental conditions. A quantitative description has been proposed by Helfrich [14] in the so-called Helfrich expression, that is

$$\gamma(H,K) = \gamma_0 + 2k_c(H - H_0)^2 + \bar{k}_c K \tag{8}$$

where γ_0 is a constant with respect to curvature. Eq. (8) defines the three bending elasticity constants k_c (bending rigidity), H_0 is the spontaneous curvature, and $\overline{k}_c$ is the saddle-splay constant. The total free energy of a small bilayer system with aggregation number N can be obtained by integrating Eq. (1) over the entire interfacial area A, giving

$$E_N = \gamma_0 A + 2k_c \int (H - H_0)^2 dA + \overline{k}_c \int K dA \tag{9}$$

The Helfrich expression introduces three quantities related to different aspects of bending a surfactant-lipid monolayer, that is k_c, H_0, and $\overline{k}_c$. The three quantities are, in principle, possible to determine from experiments or from detailed model calculations [15–18]. In these models, it is of essential importance that the hydrophobic tails in a surfactant/amphiphilic lipid monolayer are subjected to geometrical packing constraints that relate the area per amphiphilic molecule at the hydrophobic-hydrophilic interface (a) with the thickness of the hydrophobic part of the monolayer (ξ) and the molecular volume of the hydrophobic part (v). Geometrical packing constraints are taken into account by the following relation [19]

$$\frac{1}{a} = \frac{\xi}{v}\left(1 - \xi H + \frac{\xi^2}{3}K\right) \tag{10}$$

The surfactant/amphiphilic lipid monolayers are usually treated as an incompressible medium with a constant molecular volume. Notably, the geometrical relation in Eq. (10) is exact within a second-order expansion in curvature and, as a result, the Helfrich approach is expected to be accurate for aggregates with comparatively high interfacial curvature.

It has previously been demonstrated that a rather abrupt transition from micelles to bilayer aggregates is predicted to occur according to the following rather simple expression [12, 15]

$$H_0 = \frac{k_c H_0}{k_c} = \frac{1}{4\xi} \tag{11}$$

In accordance with Eq. (11), the location of the micelle-to-bilayer transition entirely depends on the spontaneous curvature H_0, as defined in the Helfrich expression in Eq. (8) and, as a consequence, micelles in a surfactant/amphiphilic lipid mixture are expected to predominate as $H_0 > 1/4\xi$, whereas bilayer aggregates predominate as $H_0 < 1/4\xi$.

Previous model calculations have demonstrated that the product $k_c H_0$ is more readily interpreted from a physical point of view than H_0 itself [15–17, 20]. Below, $k_c H_0$ denotes the effective spontaneous curvature, the quantity of which depends on molecular properties such as head group charge number, tail structure, and the hydrophilic-lipophilic balance (HLB) in a straightforward way. Hence, we are able to conclude, in accordance with Eq. (11), that micelles are favoured by high effective spontaneous curvatures $k_c H_0$ and low-bending rigidities k_c, whereas bilayer aggregates are favoured by high values of k_c and low values of $k_c H_0$.

4. Geometrically closed unilamellar bilayer vesicles

A spherical unilamellar vesicle consists of two distinct geometrical parts, a positively curved outer monolayer and a negatively curved inner layer, both with approximately same thickness ξ (*cf.* **Figure 1**). Denoting the radial distance that separates the two monolayers R_v, the hydrocarbon-water contact interface is located at a radial distance $R_e = R_v + \xi$ and the corresponding interface of the inner monolayer is located at $R_i = R_v - \xi$. The free energy of forming a unilamellar vesicle out of free monomers in solution is obtained as a sum of contributions from the outer and inner monolayers, respectively [21–23]. Since the spherical interfaces of the two monolayers each have a homogeneous curvature, the free energy of a unilamellar vesicle can be simply evaluated as

$$\mathrm{E}_{ves} = A_e\gamma_e + A_e\gamma_e = 4\pi\left(R_e^2\gamma_e + R_i^2\gamma_i\right) \tag{12}$$

γ_e and γ_i are obtained from Eq. (8) by inserting the proper expressions, $H = 1/R_e$ and $K = 1/R_e^2$ for the outer layer and $H = -1/R_i$ and $K = 1/R_i^2$ for the inner layer, that is

$$\gamma_e = \gamma_0 + 2k_c\left(\frac{1}{R_e} - H_0\right)^2 + \frac{\overline{k}_c}{R_e^2} \tag{13}$$

$$\gamma_i = \gamma_0 + 2k_c\left(-\frac{1}{R_i} - H_0\right)^2 + \frac{\overline{k}_c}{R_i^2} \tag{14}$$

where $\gamma_0 = \gamma_p - 2k_cH_0^2$ and γ_p is the interfacial tension of a planar bilayer. Introducing the dimensionless vesicle radius defined as $r_v = R_v/\xi$, we may deduce the following expression from Eq. (12)

$$\frac{\mathrm{E}_{ves}}{kT} = \alpha_v + 8\pi\lambda r_v^2 \tag{15}$$

where the dimensionless parameter

$$\alpha_v = \frac{4\pi k_{bi}}{kT} \tag{16}$$

corresponds to the work of bending a planar bilayer into a geometrically closed bilayer vesicle (bending energy) with identical interfacial area. The bilayer bending constant is defined as

$$k_{bi} = 2\left(2k_c + \overline{k}_c - 4k_c\xi H_0\right) \tag{17}$$

Although vesicles are geometrically heterogeneous aggregates with two distinct geometrical parts, that is the positively curved outer layer and the negatively curved inner layer, the bending energy in Eq. (15) consists of one single term, equal to α_v, and does not depend on vesicle size. The reason for this is that one term in each expression for the outer and inner layer, respectively, cancels out in the derivation of Eq. (15).

The second term in Eq. (15) corresponds to the work of forming a planar bilayer with an identical interfacial area as the vesicle (stretching energy), where we have introduced the dimensionless interfacial tension λ defined as

$$\lambda = \frac{\xi^2 \gamma_p}{kT} \tag{18}$$

In accordance with Eq. (7), the volume fraction of molecules aggregated in bilayer vesicles equals

$$\phi_{ves} = \frac{16\pi\xi^3}{v} e^{-\alpha_v} \int_1^\infty r_v e^{-8\pi\lambda r_v^2} dr_v = \frac{\xi^3}{v\lambda} e^{-\alpha_v - 8\pi\lambda} \tag{19}$$

and, as a consequence, the average vesicle radius equals

$$\langle r_v \rangle = \frac{\int_1^\infty r_v^2 e^{-8\pi\lambda r_v^2} dr_v}{\int_1^\infty r_v e^{-8\pi\lambda r_v^2} dr_v} = \frac{e^{-8\pi\lambda}}{2\pi\sqrt{2\lambda}\left(1 - \mathrm{erf}\sqrt{8\pi\lambda}\right)} \tag{20}$$

In the limit $\lambda \to 0$ (corresponding to $r_v \gg 1$), Eq. (19) can be rearranged so as to relate the dimensionless interfacial tension λ with the volume fraction of vesicles

$$\lambda = \frac{\xi^3 e^{-\alpha_v}}{v\phi_{ves}} \tag{21}$$

and Eq. (20) becomes simplified to

$$\langle r_v \rangle = \frac{1}{2\pi\sqrt{2\lambda}} \tag{22}$$

Combining Eqs. (21) and (22) give the following rather simple relation between average reduced vesicle radius, bending energy, and vesicle volume fraction, respectively.

$$\langle r_v \rangle = \frac{e^{\alpha_v/2}}{2\pi} \sqrt{\frac{v\phi_{ves}}{2\xi^3}} \tag{23}$$

This means that thermodynamically stable unilamellar vesicles with a finite size are predicted to exist from our theory. The average radius in Eq. (23) may be rationalized as the result of a balance between the entropy of self-assembly, tending to decrease the size of vesicles, and a size-independent positive bending energy, tending to increase the size of vesicles. Notably, the size-independent curvature energy of unilamellar vesicles is crucial for enabling the reversible formation of vesicles with finite size.

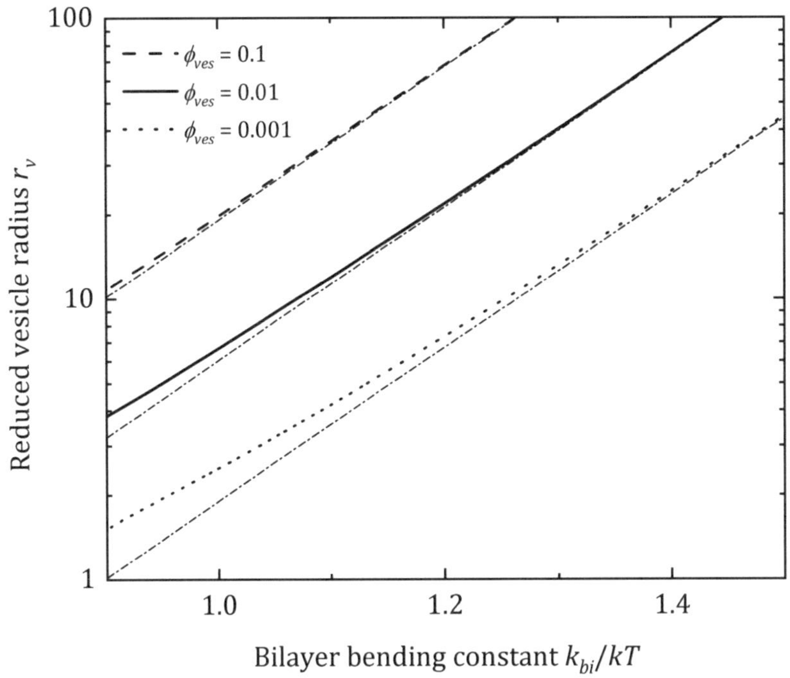

Figure 2.
Reduced vesicle midplane radius $r_v = R_v/\xi$ plotted against the bilayer bending constant k_{bi} at volume fractions ϕ_{ves} = 0.001 (dotted line), 0.01 (solid line), and 0.1 (dashed line) of vesicle bilayers in accordance with Eqs. (19) and (20). Dash-dotted lines are calculated from the approximate equation (23).

In **Figure 2**, we have plotted the average vesicle radius r_v against the bilayer bending constant k_{bi} according to Eq. (20) at some given volume fractions of vesicles. As a result of the entropy becoming more important with decreasing ϕ_{ves}, the size of the vesicles increases with increasing concentration, similar to the growth behaviour of surfactant micelles [3]. The impact of bending properties on the size of equilibrium vesicles is significant. Vesicles in the range r_v = 2–100 in dilute solutions are obtained in a rather narrow span of k_{bi} around kT, and the vesicle size increases with increasing vesicle bending energy $\alpha_v = 4\pi k_{bi}/kT$.

5. Comparison with geometrically open bilayer disks

The following expression for the curvature free energy of a circular disk with a semi-toroidal rim (*cf.* **Figure 3**) as a function of the dimensionless radius may be derived from Eq. (9)

$$E_{disk}(r_d) = \alpha_d + \pi\beta r_d + 2\pi\lambda r_d^2 \quad (24)$$

The dimensionless disk radius is defined as $r_d = R_d/\xi$, where R_d is the radius of the central planar bilayer part of the disks and ξ, as before, is half the bilayer thickness (*cf.* **Figure 3**). In Eq. (24), we have introduced the two dimensionless parameters

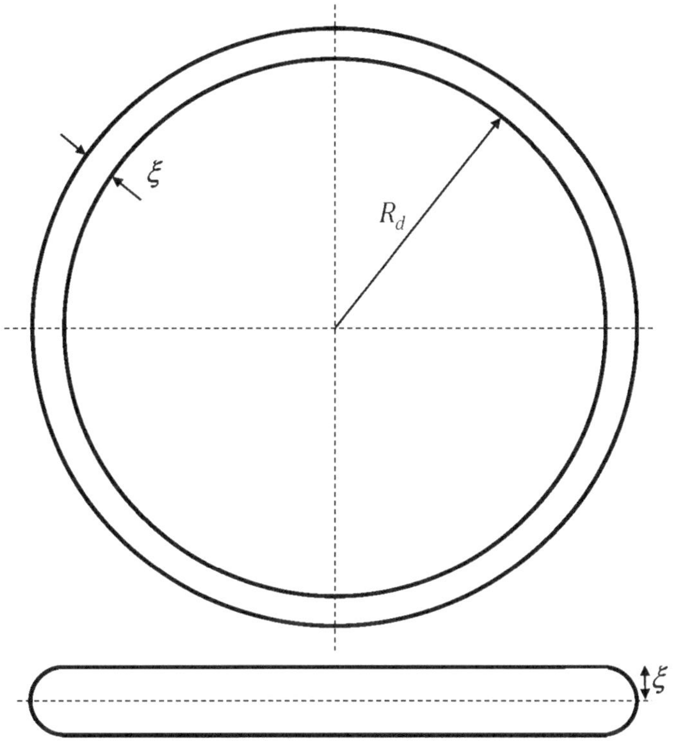

Figure 3.
Schematic picture of a disk consisting of a planar circular bilayer, with radius Rd and thickness 2ξ, surrounded by a semi-toroidal rim vid radius ξ. The upper image shows the disk when the bilayer is parallel to the paper plane. The lower image is seen from the edge side of the disk when the bilayer is perpendicular to the paper plane.

$$\alpha_d = \frac{2\pi}{kT}\left(3k_c + 2\bar{k}_c - 8\xi k_c H_0\right) \tag{25}$$

$$\beta = \frac{\pi}{kT}\left(k_c - 4\xi k_c H_0\right) \tag{26}$$

that reflect two terms with different size dependence in the expression for the curvature energy of disks. Both α_d and β depend on the three bending elasticity constants defined in the Helfrich expression in Eq. (8).

According to Eqs. (11) and (26), a transition from micelles to bilayers is expected as $\beta = 0$. Negative β-values mean that micelles are more favourable than bilayers and, hence, various bilayer structures, including disks, are expected to form as $\beta > 0$.

The total volume fraction of surfactants present as disks can be written as

$$\phi_{disk} = \frac{\pi\xi^3}{v} e^{-\alpha_d} \int_1^\infty (4r_d + 1) e^{-\left(\pi\beta r_d + 2\pi\lambda r_d^2\right)} dr_d \tag{27}$$

In accordance with Eq. (27), the formation of bilayer disks is limited as a maximum value of the volume fraction of disks is reached in the limit $\lambda \to 0$ equalling

$$\phi_{disk} = \frac{\xi^3}{v\beta}\left(5 + \frac{4}{\pi\beta}\right)e^{-\alpha_d - \pi\beta} \tag{28}$$

Hence, it follows that an appreciable amount of disks is only predicted to form close to the micelle-to-bilayer transition as the parameter $\beta \approx 0$ and Eq. (28) may be simplified to

$$\beta = \frac{2\xi^{3/2}}{\sqrt{v\pi\phi_{disk}}}e^{-\pi\left(k_c + 2\bar{k}_c\right)/kT} \tag{29}$$

Furthermore, β may assume low values only as $k_c + 2\bar{k}_c$ is larger than about unity. Likewise, the average dimensionless disk radius is obtained from the expression

$$\langle r_d \rangle = \frac{\int_1^\infty r_d(4r_d + 1)e^{-\pi\left(\beta r_d + 2\lambda r_d^2\right)}dr_d}{\int_1^\infty (4r_d + 1)e^{-\pi\left(\beta r_d + 2\lambda r_d^2\right)}dr_d} \tag{30}$$

In the limit $\lambda \to 0$, Eq. (30) turns into

$$\langle r_d \rangle = \frac{5\pi^2\beta^2 + 9\pi\beta + 8}{5\pi^2\beta^2 + 4\pi\beta} = 1 + \frac{5\pi\beta + 8}{\pi\beta(5\pi\beta + 4)} = 1 + \frac{2}{\pi\beta} - \frac{5}{5\pi\beta + 4} \tag{31}$$

It follows from Eq. (31) that appreciably large disk radii, that is $\langle r_d \rangle \gg 1$, are only possible close to the micelle-to-bilayer transition ($\beta \approx 0$) as $k_c + 2\bar{k}_c \gtrsim 1$. $\langle r_d \rangle$ larger than about unity cannot be attained for positive values of β, that is in the regime where bilayer aggregates predominate over micelles. The term including β in Eq. (24) stems from the curvature of the semi-toroidal rim, and positive values mean that the rim is unfavourable as compared to the central planar part of the disks. Even in the limit $\lambda \to 0$, the disks are prevented to grow to a finite size of appreciable magnitude due to the unfavourable rim energy.

Both the unilamellar vesicle and the bilayer disk are geometrically heterogeneous aggregates composed of different geometrical parts. However, because of the detailed geometry of the two bilayer structures, they are expected to behave completely differently from a thermodynamic point of view. The two geometrical parts of a vesicle, the positively curved outer layer and the negatively curved inner layer, together give rise to a size-independent bending energy that enables the formation of an equilibrium distribution of unilamellar vesicles of finite size in so far $H_0 < 1/4\xi$. On the other hand, the geometrically heterogeneous structure of a bilayer disk, with a central planar bilayer part and a half toroidal rim, gives rise to two curvature free energy terms that scale differently with respect to aggregate size. The formation of bilayer disks of reasonable size is only promoted by small β values (corresponding to $H_0 \approx 1/4\xi$), but so are micelles over bilayer aggregates since a transition from bilayers to micelles is expected as β turns to negative values. This means that in contrast to unilamellar vesicles, the geometrically composed structure of bilayer disks prevents the formation of large disks in the regime where bilayers are favoured over micelles.

6. Influence of bending elasticity on the size of unilamellar vesicles

According to Eqs (16) and (17), the bilayer bending constant k_{bi} and bending energy α_v increase with increasing bending rigidity k_c, increasing saddle-splay constant $\bar{k}_c$, and decreasing effective spontaneous curvature k_cH_0. Moreover, according to Eq. (23), the size of equilibrated vesicles follows similar trends with the three bending elasticity constants. This means that a small size of unilamellar vesicles is favoured by small values of k_c and $\bar{k}_c$, and high values of k_cH_0.

The spontaneous curvature (H_0) represents the sign and magnitude of the preferential curvature of a single surfactant monolayer. The dependence of k_cH_0 on the size of unilamellar vesicles is the result of the outer vesicle monolayer always being more voluminous than the inner layer with a larger number of surfactant/lipid molecules. This tendency increases in magnitude with decreasing vesicle size. A small vesicle size means larger curvature of the outer vesicle layer and smaller (i.e. more negative) curvature of the inner layer, and since effects in the outer layer trump effects in the inner layer, vesicles decrease in size with increasing k_cH_0.

The bending rigidity (k_c) is a measure of the ability of a monolayer to resist deviations from a uniform mean curvature equal to the spontaneous curvature. k_c must always be a positive quantity in order to realize a minimum of γ as a function of mean curvature in Eq. (8). Large bending rigidities favour geometrically homogeneous aggregates with a uniform curvature, or in the case of geometrically heterogeneous aggregates, smaller deviations in curvature between the different geometrical parts. This means that k_c usually mainly influences the shape, but not the size, of an aggregate, which is true, for instance, for geometrically homogeneous spherical microemulsion droplets. However, in the case of a unilamellar vesicle, with a positively curved outer layer and a negatively curved inner layer, the difference in curvature between the two geometrical parts decreases in magnitude with increasing vesicle size. This means that, in contrast to micelles and microemulsion droplets, the bending rigidity has an explicit influence on the size of unilamellar vesicles. Hence, small vesicles are favoured by low values of k_c.

The third bending elasticity constant, the saddle-splay constant ($\bar{k}_c$), is related to the Gaussian curvature K. As implied by its name, high positive values of $\bar{k}_c$ influence the curvature of an interface to favour a saddle-like structure, with negative Gaussian curvature, that is curvatures with opposite signs in perpendicular directions. According to the Gauss-Bonnet theorem, the last integral in Eq. (9) is always equal

$$\int KdA = 4\pi(1-g) \tag{32}$$

where the genus g represents the number of handles or holes present in a surfactant monolayer. From a geometrical point of view, a vesicle consists of two closed interfaces (i.e. the inner and outer monolayers, respectively) giving $g = -1$. As a result, Eq. (32) equals $8\pi\bar{k}_c$, the quantity of which does not depend on the size of the vesicle. Hence, from a mathematical perspective, $\bar{k}_c$ has a similar impact as k_cH_0 and k_c, contributing a size-independent term to k_{bi} and α_v. Consequently, small vesicles are favoured by small (possibly negative) values of $\bar{k}_c$. We may also note that vesicles belong to a different topology ($g = -1$) than micelles ($g = 0$). This means that micelles, with a higher genus number, are favoured with respect to vesicles as $\bar{k}_c$ assumes positive values, whereas the opposite holds (vesicles favoured with respect to micelles) when $\bar{k}_c$ is negative. This effect is taken into account in the simple relation in Eq. (11).

7. Effects of molecular structure on the bending elasticity constants

Whether unilamellar vesicles may form spontaneously or not, or the size of equilibrium vesicles, largely depends on the chemical structure of the amphiphilic components making up a bilayer. It also depends on environmental quantities such as ionic strength, pH, and temperature. Hence, the size of spontaneous unilamellar vesicles may be rationalized by means of investigating the molecular interpretation of the various bending elasticity constants.

The main influence from the head group of amphiphilic molecules comes from the molecular charge number, which gives rise to a surface charge in the head group layers in a bilayer. The effect of surface charge density occurring from ionic surfactants and lipids on various bending elasticity constants may be calculated from the Gouy-Chapman theory [24]. Electrostatic effects tend to raise both spontaneous curvature and bending rigidity while contributing a negative value to the saddle-splay constant. Bilayer aggregates are expected to predominate over micelles at lower values of the spontaneous curvature according to Eq. (11). This means that the presence of smaller vesicles is usually promoted by small values of k_c, rather than large H_0, and lower surface charge densities.

The head group of a nonionic surfactant appears to have a similar effect of promoting high and positive values of the effective spontaneous curvature k_cH_0 [18]. The effect on bending rigidity appears to be more complicated and it has been theoretically demonstrated that a maximum k_c appears at a certain hydrophilic-lipophilic balance (HLB) of a nonionic surfactant [18].

It has been demonstrated that additional contributions to all bending elasticity constants, due to the finite thickness of the hydrophobic part, appear for a monolayer composed of surfactant or amphiphilic lipid with flexible hydrocarbon tails [16]. Most interestingly, this contribution is completely absent for molecules with a rigid tail. The hydrophobic finite thickness effects tend to substantially lower the effective spontaneous curvature k_cH_0 as well as increase the bending rigidity k_c. Lowering k_cH_0 tends to favour bilayer aggregates over micelles and a flexible hydrocarbon tail of at least one component seems to be necessary for a bilayer to form at all. However, the finite thickness effects also tend to raise k_{bi} and the vesicle bending energy, thus promoting larger vesicles rather than smaller vesicles. The flexibility of the hydrophobic part may be reduced by adding one or more double bonds to an aliphatic chain. Most interestingly, conspicuously small (ultrasmall) unilamellar vesicles have been observed to form spontaneously when a drug surfactant is mixed with a phosphatidyl choline phospholipid with unsaturated hydrophobic tail, but not with a phospholipid with a saturated tail [11].

8. Effect of mixing amphiphilic molecules

The spontaneous formation of unilamellar vesicles has virtually exclusively been observed in surfactant/surfactant or surfactant/lipid mixtures. Most interestingly, when investigating the three bending elasticity constants k_cH_0, k_c, and $\bar{k}_c$, it turns out that the bending rigidity behaves fundamentally differently than the other two constants in the sense that k_c has an explicit contribution from the effect of mixing two amphiphilic molecules that always brings it down [17, 20, 25, 26]. The effect is enhanced as the asymmetry between two components is increased and k_c assumes a minimum magnitude as the composition in the aggregates is optimized. By means of reducing k_c while leaving k_cH_0 approximately unchanged, it is possible to promote the formation of small unilamellar vesicles without promoting micelle formation.

Figure 4 shows model calculations of the mixing contribution to the bending rigidity k_c for mixtures of an ionic and nonionic surfactant with identical flexible hydrophobic tails [17]. It is seen that pure mixing effects might significantly bring down k_c, and there exists an optimized composition in the bilayer aggregates where k_c is expected to reach a minimum value. It has been demonstrated that an ionic single-tailed/nonionic double-tailed surfactant mixture is more asymmetric than an ionic single-tailed/nonionic single-tailed surfactant mixture, giving rise to an even more negative mixing contribution to k_c [17].

The reason for the reduction of bending rigidity, and as a consequence, the bilayer bending constant as well as the average size of vesicles, is that the composition in a monolayer consisting of two surfactants with different spontaneous curvatures becomes a strong function of curvature. As a result, different equilibrium compositions in the outer and inner monolayers of a vesicle, where the amphiphilic component with a high spontaneous curvature prefers the outer layer and the component with lower H_0 prefers the inner layer, give rise to lower bending rigidity and smaller vesicles (*cf.* **Figure 5**).

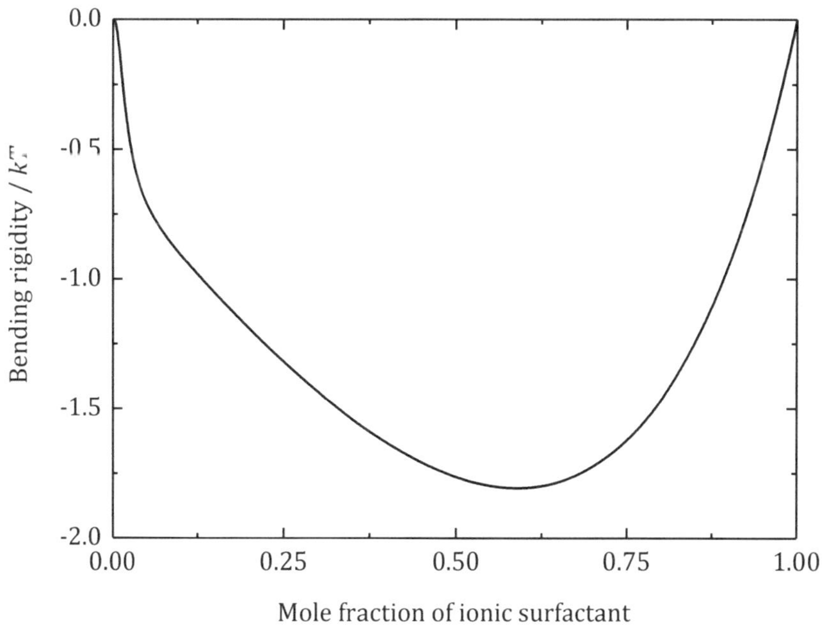

Figure 4.
Explicit mixing contribution to the bending rigidity k_c for a mixture of an ionic and a nonionic surfactant with identical tails [17].

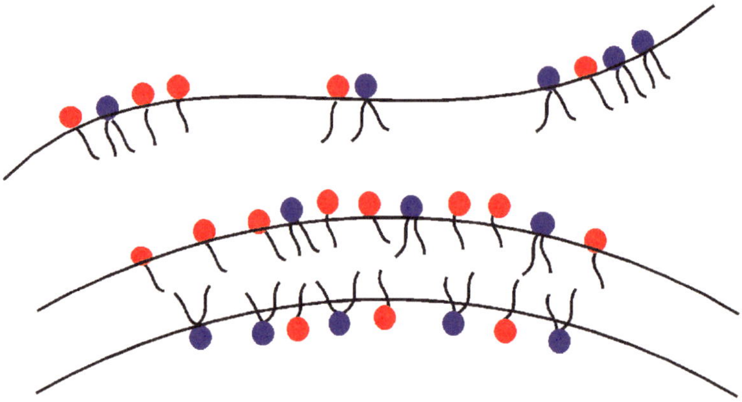

Figure 5.
Schematic images of a monolayer and a bilayer composed of two amphiphilic components. The component with the red head group prefers to be located in parts with a positive curvature and the component with the blue head group to be located in parts with negative curvature [17].

Spontaneous formation of small unilamellar vesicles has been observed in several aqueous systems of two oppositely charged surfactants, where one of the surfactants is present in excess. In this case, the asymmetry between the surfactants comes from the different head group charge numbers, giving rise to a difference in charge density between the two monolayers. The surfactant in excess prefers the outer layer, implying a higher charge density in the outer positively curved layer and a lower charge density in the inner layer that is preferred by the surfactant in deficit. The difference in charge number between two surfactants with z = +1 and −1 is two, which enhances the reduction in k_c and k_{bi} as compared with a mixture of a monovalent ionic surfactant and a nonionic surfactant [17]. Moreover, mixing an ionic single-tailed surfactant with a double-tailed oppositely charged surfactant, in excess of the single-tailed surfactant, enhances the asymmetry between the surfactants and, as a consequence, the reduction of k_c is enhanced. As a matter of fact, ultrasmall unilamellar vesicles have been observed in mixtures of the ionic single-tailed surfactant sodium dodecyl sulphate (SDS) with the double-tailed cationic surfactant didodecyldimethyl ammonium bromide (DDAB) dissolved in water in excess of SDS [27].

Ultrasmall vesicles, with a diameter less than 20 nm, have also recently been observed to form spontaneously in mixtures of an ionic surfactant with short and rigid tail and a zwitterionic phospholipid at high ionic strength [11]. In this case, there is a large asymmetry in spontaneous curvature with the small surfactant preferring to be located in the outer layer and the phospholipid preferring the inner layer of the vesicle. In a dilution series, the vesicles are found to reach a minimum size at some optimal surfactant/lipid composition.

Similar to a vesicle, a bilayer disk is a geometrically heterogeneous structure composed of a central planar bilayer surrounded by a more curved semi-toroidal rim at the edge of the disk. In a binary mixture of amphiphilic components, the component with higher spontaneous curvature prefers to be located in the rim, whereas the other component prefers to be located in the planar central part. As for vesicles, the segregation of components has an explicit impact on bending rigidity k_c but not on the other bending elasticity constants. Lowering k_c by means of mixing amphiphilic components will reduce the bending parameter α_d in Eq. (25), but this effect cannot promote the formation of large disks, since the formation of the latter is restricted to the vicinity of the micelle-to-bilayer transition as $\beta \approx 0$.

9. Concentrated vesicle solutions and overpacking

The theoretical arguments presented so far are strictly valid for a dilute solution of vesicles where inter-aggregate interactions have been neglected. Since vesicles, for geometrical reasons, are comparatively voluminous structures, including a substantial amount of aqueous phase in the interior of the closed vesicle bilayer, they may become overpacked above a certain concentration of total amphiphilic component. This limit is significantly reduced as the size of the vesicles increases. Assuming a simple cubic packing of the vesicles, the overpacking limit, in terms of the solute volume fraction where spherical vesicles are closely packed, can be calculated from the following simple relation

$$\phi_{pack} = \frac{8\pi R_v^2 \xi}{(2R_v)^3} = \frac{\pi}{r_v} \tag{33}$$

where $r_v = R_v/\xi$. Hence, Eq. (33) reveals the overpacking limit to be inversely proportional to the vesicle radius. This means that comparatively large vesicles might be overpacked at very low surfactant/lipid concentrations. Moreover, inter-aggregate interactions may play a significant role even at substantially lower concentrations than ϕ_{pack}. A consequence of these interactions is that geometrically closed vesicles open up to form bilayer disks that may pack more densely by means of excluding the water trapped in the vesicle cores. According to Eq. (33), the overpacking limit for substantially large vesicles, that is r_v larger than 100, is lower than a few volumes per cents. This is consistent with the experimental observations that a significant amount of disks are usually present in solutions of comparatively large vesicles and the amount of disks seems to increase with increasing vesicle size [7, 27, 28].

10. Conclusions

Liposomes shaped as unilamellar vesicles may form spontaneously in certain surfactant-phospholipid mixtures and it is possible to manufacture a novel type of thermodynamically stable colloidal solution with similar properties as micellar and microemulsion solutions [11]. This kind of solution consists of rather small vesicles, the size of which is stable over time in a fixed environment and the solution may become reversibly reproduced when returning to its original environmental conditions. We have demonstrated the theoretical principles of colloidal liposome solutions by combining bending elasticity theory and solution thermodynamics. In particular, we have shown that geometrically closed bilayer vesicles may be equilibrium structures with a wide range of finite average sizes. The work of bending a planar bilayer into a geometrically closed vesicle is constant with vesicle size. As a result, equilibrium vesicles with finite size may form as a result of a balance between positive bending energy, favouring larger vesicles, and entropy of self-assembly favouring smaller vesicles. Hence, the average size of equilibrated vesicles increases with increasing bending energy (or bilayer bending constant) as well as increasing vesicle bilayer concentration. Moreover, bilayer aggregates such as vesicles and disks are expected to predominate over micelles in so far the spontaneous curvature falls below $H_0 = 1/4\xi$, where ξ denotes the monolayer thickness. Small unilamellar vesicles are favoured by small values of the bilayer bending constant $k_{bi} = 2(2k_c + \bar{k}_c - 4k_c\xi H_0)$. k_{bi} may be reduced by means of increasing the spontaneous curvature. However, increasing H_0 favours micelles over bilayer aggregates. A more efficient way to reduce k_{bi} to magnitudes compatible with small unilamellar vesicles, while still being in the region where bilayers predominate over micelles, is to reduce the bending rigidity k_c. Most interestingly, k_c may be reduced by means of mixing two amphiphilic components with different spontaneous curvatures, where one of the components prefers the outer, oppositely curved, monolayer, whereas the other prefers the inner, negatively curved, layer.

In contrast to vesicles, bilayer lipodisks may only be stable under exceptional circumstances, such as near a micelle-bilayer coexistence region or at amphiphile concentrations close to or above the overpacking limit of vesicles. The positively curved rim of disks cannot be stable in a regime where bilayers are favoured over micelles by low spontaneous curvature of its components. Reducing bending rigidity by means of mixing asymmetric components has no impact on the thermodynamic stability of non-interacting disks. However, disks may become more stable at higher concentrations, above the overpacking limit of unilamellar vesicles, because disks are able to pack more densely than vesicles. Since the overpacking limit concentration depends on vesicle size (it is proportional to one over vesicle radius), disks are expected to be favoured by large values of the bilayer bending constant k_{bi}, large bilayer aggregate sizes, and high concentrations of amphiphilic components.

Author details

Lars Magnus Bergström
Department of Medicinal Chemistry, Uppsala University, Sweden

*Address all correspondence to: magnus.bergstrom@ilk.uu.se

References

[1] Bergström LM. Thermodynamics of self-assembly. In: Tadashi M, editor. Application of Thermodynamics to Biological and Material Science. London, UK, Rijeka: IntechOpen; 2011. 289-314 pp. DOI: 10.5772/1763

[2] Bergström LM. Thermodynamics and bending energetics of microemulsions. In: Karunaratne DN, editor. Properties and Uses of Microemulsions. London, UK: IntechOpen; 2017. 117-137 pp. DOI: 10.5772/67369

[3] Bergström LM. Explaining the growth behavior of surfactant micelles. Journal of Colloid and Interface Science. 2015; **440**:109-118. DOI: 10.1016/j.jcis.2014.10.054

[4] Lichtenberg D. Characterization of the solubilization of lipid bilayers by surfactants. Biochimica et Biophysica Acta. 1985;**821**:470-478. DOI: 10.1016/0005-2736(85)90052-5

[5] Tondre C, Caillet C. Properties of the amphiphilic films in mixed cationic/anionic vesicles: A comprehensive view from a literature analysis. Advances in Colloid and Interface Science. 2001;**93**: 115-134. DOI: 10.1016/S0001-8686(00)00081-6

[6] Bergström M, Pedersen JS, Schurtenberger P, Egelhaaf SU. Small-angle neutron scattering (SANS) study of vesicles and lamellar sheets formed from mixtures of an anionic and a cationic surfactant. The Journal of Physical Chemistry. B. 1999;**103**: 9888-9897. DOI: 10.1021/jp991846w

[7] Bergström LM, Skoglund S, Edwards K, Eriksson J, Grillo I. Self-assembly in mixtures of an anionic and a cationic surfactant: A comparison between small-angle neutron scattering and cryo-transmission electron microscopy. Langmuir. 2013;**29**: 11834-11848. DOI: 10.1021/la401884k

[8] Bergström LM, Skoglund S, Eriksson J, Edwards K, Grillo I. Spontaneous transformations between surfactant bilayers of different topologies observed in mixtures of sodium octyl sulfate and hexadecyltrimethylammonium bromide. Langmuir. 2014;**30**:3928-3938. DOI: 10.1021/la4042259

[9] Egelhaaf SU, Schurtenberger P. Shape transformations in the Llecithin-bile salt system: From cylinders to vesicles. The Journal of Physical Chemistry. 1994;**98**: 8560-8573. DOI: 10.1021/j100085a041

[10] Pedersen JS, Egelhaaf SU, Schurtenberger P. Formation of polymerlike mixed micelles and vesicles in lecithin-bile salt solutions: A small-angle neutron scattering study. The Journal of Physical Chemistry. 1995;**99**: 1299-1305. DOI: 10.1021/j100004a033

[11] Forooqi Motlaq V, Gedda L, Edwards K, Doutch J, Bergström LM. Spontaneous formation of ultrasmall unilamellar vesicles in mixtures of an amphiphilic drug and a phospholipid. Langmuir. 2023;**39**:11337-11344. DOI: 10.1021/acs.langmuir.3c01023

[12] Bergström LM. Bending energetics of tablet-shaped micelles: A novel approach to rationalize micellar systems. ChemPhysChem. 2007;**8**:462-472. DOI: 10.1002/cphc.200600692

[13] Eriksson JC, Ljunggren S, Henriksson U. A novel approach to the mechanics and thermodynmaics of spherical micelles. Journal of the Chemical Society, Faraday Transactions.

1985;**2**(81):833-868. DOI: 10.1039/F29858100833

[14] Helfrich W. Elastic properties of lipid bilayers: Theory and possible experiments. Zeitschrift für Naturforschung. Section C. 1973;**28**: 693-703. DOI: 10.1515/znc-1973-11-1209

[15] Bergström LM. Model calculations of the spontaneous curvature, mean and Gaussian bending constants for a thermodynamically open surfactant film. Journal of Colloid and Interface Science. 2006;**293**:181-193. DOI: 10.1016/j.jcis.2005.06.050

[16] Bergström LM. Bending elasticity of charged surfactant layers: The effect of layer thickness. Langmuir. 2006;**22**: 3678-3691. DOI: 10.1021/la052800z

[17] Bergström LM. Bending elasticity of charged surfactant layers: The effect of mixing. Langmuir. 2006;**22**:6796-6813. DOI: 10.1021/la060520t

[18] Bergström LM. Bending elasticity of nonionic surfactant layers. Langmuir. 2009;**25**:1949-1960. DOI: 10.1021/la802532n

[19] Israelachvili JN, Mitchell DJ, Ninham BW. Theory of self-assembly of hydrocarbon amphiphiles into micelles and bilayers. Journal of the Chemical Society, Faraday Transactions. 1976; **2**(72):1525-1568. DOI: 10.1039/F29767201525

[20] Porte G, Ligoure C. Mixed amphiphilic bilayers: Bending elasticity and formation of vesicles. The Journal of Chemical Physics. 1995;**102**:4290-4298. DOI: 10.1063/1.469476

[21] Bergström M, Eriksson JC. The energetics of forming equilibrated bilayer vesicles. Langmuir. 1996;**12**: 624-634. DOI: 10.1021/la950282u

[22] Bergström M. Thermodynamics of vesicle formation from a mixture of anionic and cationic surfactants. Langmuir. 1996;**12**:2454-2463. DOI: doi.org/10.1021/la950873k

[23] Bergström M. Thermodynamics of unilamellar vesicles: Influence of mixing on the curvature free energy of a vesicle bilayer. Journal of Colloid and Interface Science. 2001;**240**:294-306. DOI: 10.1006/jcis.2001.7624

[24] Mitchell DJ, Ninham BW. Curvature elasticity of charged membranes. Langmuir. 1989;**5**:1121-1123. DOI: 10.1021/la00088a044

[25] Safran SA, Pincus P, Andelman D. Theory of spontaneous vesicle formation in surfactant mixtures. Science. 1990; **248**:354. DOI: 10.1126/science.248.4953.354

[26] Kozlov MM, Helfrich W. Effects of a cosurfactant on the stretching and bending elasticities of a surfactant monolayer. Langmuir. 1992;**8**:2792-2797. DOI: 10.1021/la00047a035

[27] Bergström M, Pedersen JS. A small-angle neutron scattering study of surfactant aggregates formed in aqueous mixtures of sodium dodecyl sulfate and didodecyldimethylammonium bromide. The Journal of Physical Chemistry. B. 2000;**104**:4155-4163. DOI: 10.1021/jp993666s

[28] Bergström LM, Skoglund S, Danerlöv K, Garamus VM, Pedersen JS. The growth of micelles, and the transition to bilayers, in mixtures of a single-chain and a double-chain cationic surfactant investigated with small-angle neutron scattering. Soft Matter. 2011;**7**: 10935-10944. DOI: 10.1039/C1SM05760J

Section 2

Liposome Applications

Chapter 3

Latest Advances in Application of Liposomes and Nanoliposomes

Ehsan Kianfar, Abolfazl Mirani, Zaid H. Mahmoud and Arezoo Mohammadkhani

Abstract

Liposomes are spherical vesicles possessing a membrane composed of a phospholipid bi-layer and used to deliver drugs or genetic material into a cell. Research on liposome technology has progressed over the past two decades in view of developing long-circulating liposomes through modulating lipid composition, size, and charge of the vesicle. The advantages of liposomes include easy preparation and packaging of the exogenous DNA. Liposomal vectors are non-viral and thus relatively nontoxic as compared to viral vectors; moreover, they can pass the blood-brain barrier. However, in this method, genes are not integrated into genome and low expression. There is no denying that nanotechnology has revolutionized the human life, enabling enlargement of materials variety and application. Due to the increased need for food, human beings revisited and optimized processes to open new corridors to enhance the quality of life. Nanoliposomes are available in different types and can be produced applying different methods such as encapsulation. The food active ingredients can be stabilized against chemical and environmental changes, preventing their decomposition and thereby increasing their storage time and accessibility.

Keywords: liposomes, nanoliposomes, nanotechnology, absorption, food

1. Introduction

Liposomes were discovered in the mid-1960s by Alec Bangham and were originally known as Banghosomes and were originally investigated as membrane models. In fact, the discovery of Liposomes was based on the fact that by mixing lipids and drugs, small vesicles containing drugs are obtained, which, due to the ease of transport and the absence of obstacles in the way of transportation of the aforementioned vesicles, the use of vesicles as a way to transport drugs. The treatment of some diseases has been investigated, and in fact, in this way, they have been able to transfer some toxic drugs that are anti-tumor or drugs that are not easily absorbed in the body to the target tissue [1, 2]. Liposomes are spherical vesicles whose membrane components are phospholipids and have a diameter of 20 nm–10 μm. Phospholipid molecules, which belong to the group of fats, consist of two hydrophilic and hydrophobic parts, their tail part is a long non-polar and hydrophobic structure and consists of two hydrocarbon chains, a saturated chain and an unsaturated chain, the head part is the polar, and hydrophilic structure consists of glycerol group, phosphate group, and choline group.

Thus, phospholipids have polar and non-polar ends, and various phospholipid molecules such as phosphatidyl acid, phosphatidylinositol, phosphatidylserine, phosphatidylcholine (lecithin), phosphatidyl glycerol, and cardiolipin can be used to make liposome. Liposomes can be single-layered or multi-layered. The selection of two-layer composition can determine the fluidity and type of liposome load [3, 4]. For example, saturated phospholipids with long acyl chains such as dipalmitoyl phosphatidylcholine cause more stiffness in the bilayer and create an impermeable structure. While unsaturated types of phosphatidylcholines from natural sources such as eggs or soybeans give liposome less stability and greater permeability. The size of the number of layers, the hardness of the two layers or their fluidity, the load and the changes on the surface of the liposome, and all these parameters determine the fate of the liposomes in the glass environment. One of the applications of this biological system is gene therapy, and in order to carry out gene therapy, we definitely need vectors or carrier systems for transferring the desired gene [5, 6]. In this method, the vectors used are divided into two types of viral and non-viral vectors, liposomes. For gene therapy, they act as non-viral vectors, which have a much lower risk than viral vectors, which is one of the issues that have attracted the attention of many researchers, and among them, cationic liposomes are used more often. Cationic liposomes are mixed with negatively charged DNA and form compounds called lipoplexes, and the interaction of liposomes with DNA and subsequently the formation of lipoplexes depends on a series of physical conditions. Due to their amphipathic properties, liposomes provide the possibility of drug delivery for hydrophilic and lipophilic drugs, so that with the use of this biological system, many obstacles on the way of drug delivery to different tissues will be removed, and a new chapter for development and new areas of progress will be removed [7]. It will be opened in medicine, including nanomedicine. Many characteristics of liposomes, such as low inherent toxicity, biodegradability, and lack of immunogenicity, have caused liposomes to be of interest to many researchers, especially scientists in the fields of biotechnology and nanomedicine, as a gene and drug delivery system. Nanomedicine is actually the use of nanotechnology in the treatment of diagnosis and control of biological systems, which includes identifying the relevant target and choosing appropriate carriers to achieve appropriate responses and minimize the side effects of drugs.

2. Preparation steps of liposomes

The discovery of liposomes was based on the fact that by mixing lipids and drugs and quickly shaking the mixture obtained by ultrasonic shaking devices, very small spheres were obtained from very small drops of fat, in which layers the water-soluble drug was placed between the layers of fat. The goal was that by injecting liposomes into the circulatory system, they would be selected and absorbed by tumor cells, and after breaking them at the site of the tumor cells, they would sprinkle the drugs on the cancer cells without harming the healthy cells of the body, to transfer [8, 9]. The characteristics of liposome synthesis can be changed depending on their composition, cationic, anionic, and neutral. Nevertheless, regardless of the composition of liposomes, the same preparation method can be used for all vesicles. In the preparation of liposomes, lipids obtained from natural sources such as eggs or soybeans are first dissolved in an organic solvent and mixed well to obtain a homogeneous mixture. In this step, a clear solution should be obtained. After this step, the organic solvent is removed from the container, and we have the lipid membranes. To ensure the complete removal of the solvent, the lipid solution is frozen by placing it on dry ice molds or rotating it in an acetone ice bath. Then the frozen lipid template is placed on a vacuum pump and lyophilized until it is completely dry. In the next step, we have the hydration of the lipid membranes, which is done by adding aqueous medium to the dried lipid membrane and then stirring it. Usually, the product of hydration consists of large multi-layered vesicles (LMV) that have a structure similar to an onion, and each layer is separated by a layer of water [10, 11]. In the next step, the lipid solution containing LMV is subjected to sonication and smaller, and homogeneous particles are produced. At this stage, small single-layer vesicles (SUV) with a diameter of about 15–50 are usually produced. Of course, in the meantime, other lipids are also formed, such as large monolayer lipids (LUV), then the suspension is checked by gel filtration, and the appropriate liposomes are isolated.

3. Components of liposomes

The building blocks of lipids are: phospholipids, cholesterol, steric stabilizers, and load generator.

3.1 Phospholipids

Phospholipids are the most important constituents of biological membranes and also form the main structural component of liposomes. Phospholipids or phosphatides are glycerides that have phosphoric acid and an alcohol in their structure. In phospholipids, the part of alcohol and phosphoric acid is polar, while the part contains non-polar fatty acids, in this sense, they are called amphipathic compounds, and because of this amphipathic characteristic, they create a double-layer membrane [12].

3.2 Cholesterol

The most widely used type of sterol in the production of lipid vesicles is cholesterol. The structure of cholesterol consists of four hydrocarbon rings called steroid rings, a hydroxyl group, and a hydrocarbon chain. Cholesterol increases the stability of liposomes in such a way that the polar hydroxyl group of cholesterol is placed near the carbonyl group of phosphatidylcholines through the formation of a hydrogen bond, and the hydrophobic hydrocarbon rings of cholesterol align themselves with the fatty acid chains of phosphatidylcholine. As a result, the bond between cholesterol and phosphatidylcholine increases the electrostatic repulsion between phospholipid layers and finally by limiting the movement of acyl phospholipid chains increases liposome stability [13].

3.3 Steric stabilizers and load generator

Sterilamine and diacetyl phosphate are the most common substances for creating positive and negative charges on the surface of liposomes. The use of these substances increases the stability and absorption of liposomes [14].

4. Classification of liposomes

The size of liposomes can be as small as 0.025 to as large as 2.5 μm. For this reason, these particles are included in the scope of nanotechnology. Also, liposomes can be single-walled or multi-walled (from two phospholipid layers). The size of liposomes is the factor that determines the half-life of these structures in the circulatory system of the body. In general, liposomes are classified based on their size and phospholipid layers (**Figure 1**) [15, 16]:

- Multilamellar vesicles or (MLV): Multi-walled liposomes with a size of more than 400 nm.
- Large unilamellar vesicles or (LUV): Single-walled liposomes with a size of about 100–400 nm.
- Small unilamellar vesicles (SUV): Single-walled liposomes with a size of about 25–100 nm.

Liposomes are vesicles characterized by a lipid bilayer that encompasses a watery center. They are primarily composed of phospholipids recognizable by a polar hydrophilic head and two a polar hydrophobic chain. When scattered in fluid arrangements, their polar heads associated with the watery environment, due to the hydrogen bonds and polar intuitive, whereas their aliphatic chains connected with each other due to the van der Waals strengths, driving to a lipid bilayers arrangement of which they constitute the lipophilic inward compartment. Amid their detailing prepare, water-soluble drugs can be broken down within the fluid compartment, whereas hydrophobic materials can be entangled into the lipid bilayer. Joined by being composed of biocompatible materials, liposomes appear diverse structures, measurements, lipid composition, and surface charge. They can be composed by a few concentric bilayers isolated by fluid compartments, with an outside lipid bilayer containing other ever littler bilayers isolated by water cavities, like an onion structure. In this case, liposomes are called multilamellar vesicles (MLVs) and appear a measure run of 500 nm–5 μm, or by as it were one phospholipid bilayer encompassing a fluid compartment. In this case, liposomes can be separated in little, expansive, and mammoth vesicles depending upon their measurement: They are called little unilamellar vesicles (SUVs) on the off chance that they have a 20–200 nm run measure, huge unilamellar vesicles (LUVs) with a 200–1 μm extend measure, and mammoth unilamellar vesicles (GUVs) with

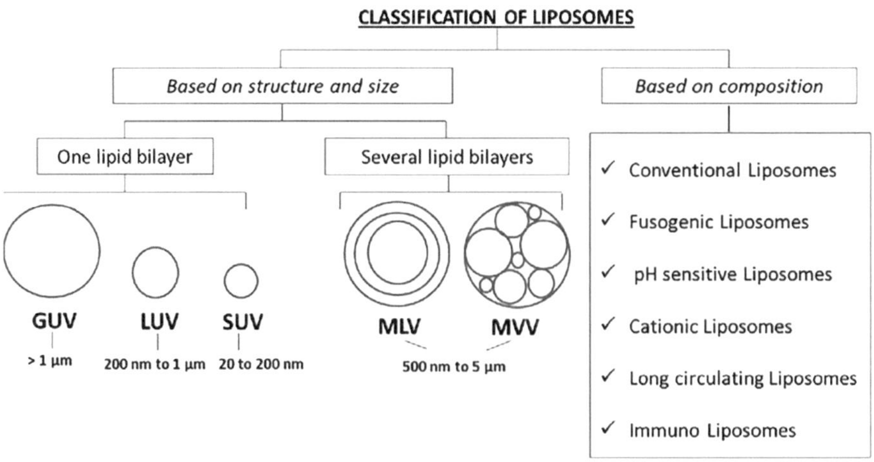

Figure 1.
Classification of liposomes based on size [1].

an estimate bigger than 1 μm. At last, comparable in measurement to MLVs there are multi-compartmental structures constituted by vesicles encompassed by other vesicles called multivesicular vesicles (MVVs).

5. Types of liposomes

Liposomes are classified into three different categories based on their nature and composition (**Figure 2**) [17, 18]:

- Stealth liposomes: Due to their surface receptors, these liposomes can avoid being phagocytosed by the cells of the immune system and remain in the body for a longer period of time. These liposomes are used for drugs that require long-term release in the body. The surface receptor used for these liposomes is usually polyethylene glycol (PEG).
- Conventional liposomes: They are among the most widely used liposomes. This group has a negative charge due to the number of phospholipids and cholesterol they have and are used to deliver drugs to different tissues.
- Targeted liposomes: Targeted liposomes can specifically deliver drugs to a specific tissue or cell. Antibodies are placed on the surface of these liposomes so that they can identify the cell in question.
- Cationic liposomes: This group of liposomes have a positive charge due to having some cations on their surface and are mostly used in gene therapy.

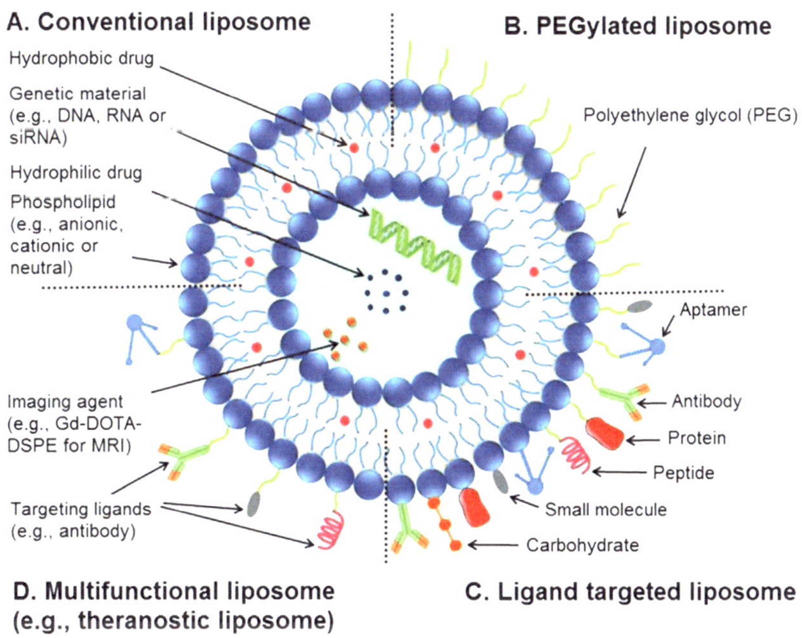

Figure 2.
Types of liposomes based on nature and function [2].

The foremost known lipids utilized within the liposomal definitions are phosphatidylcholine (zwitterionic), phosphatidylglycerol (adversely charged), phosphatidic corrosive, phosphatidylethanolamine (zwitterionic), and phosphatidylserine (contrarily charged). Emphatically charged lipids (e.g., N-[1-(2,3-dioleyloxy) propyl]-N, N,N-triethylammonium (DOTMA), and 1,2-dioleoyl-3-trimethylammoniopropane (DOTAP)) are primarily utilized for quality conveyance, as they associated with the adversely charged deoxyribonucleic corrosive (DNA) and contrarily charged APIs. Cholesterol is another vital component of liposomes. It includes a modulatory effect on the properties of the lipid bilayer of the liposomes. It can control the heftiness within the liposome structure and increment the pressing between the phospholipid particles, coming about in more requested adaptation within the aliphatic tail region, reduced small-scale extremity, decreased bilayer adaptability to the encompassing atoms (particularly water-soluble particles), and increments within the miniaturized scale thickness of the bilayer. Cholesterol is additionally significant for basic steadiness of liposomal films against intestinal natural stretch. Cholesterol was found to impact liposomes estimate (expanding cholesterol concentration increments liposomes estimate in expansion to shape move), give porousness and ease, and thus tweak the discharge of hydrophilic compounds from liposomes. It is additionally conceivable to utilize surface functionalization of liposomes by an assortment of operators to overcome the restrictions of these nanocarriers in terms of natural and physiological boundaries. For case, liposomes can be functionalized with polyethylene glycols (PEGs), aptamers, antibodies, proteins, peptides, ligands, carbohydrates, or little atoms (**Figure 2**).

6. Applications of liposomes

Liposomes have useful properties that have doubled their importance for application in many fields of cell science, medicine, and health. In this article, we will examine some of the applications of liposomes in different fields.

6.1 Gene therapy using liposomal vectors

RNA obstructions (RNAi) could be a implies of directing quality expression by siRNAs and microRNAs (miRNAs). siRNAs tie to the RNA-induced quieting complex (RISC), and once they enter the cytoplasm, they initiate quality hushing by coordinating the particular cleavage of arrangements of completely complementary matched mRNAs. Comparatively, miRNAs intervene the restraint of interpretation and the end of translation of not completely complementary mRNAs. miRNAs may to intercede mRNA debasement in cytoplasmic compartments, known as preparing bodies (P-bodies), in this way anticipating protein amalgamation. Primate synthesized siRNAs are incapable to cross organic layers by detached dissemination due to their tall atomic weight and polyanionic nature; in this manner, they require sedate conveyance procedures. Lipid nanoparticles containing siRNAs can hush qualities, and in this way, are exact helpful devices. Wanatabe's group endeavored to hush the quality expression of hepatitis C infection (HCV) in vivo utilizing novel cationic liposome-encapsulated siRNA particles. They utilized enzyme-linked immunosorbent measures (ELISA) to degree the expression of the HCV center protein in mouse liver and found that siRNA/lactosylated cationic liposome 5 (CL-LA5) complexes (siRNA/CL-LA5) particularly repressed HCV protein expression in a dose-dependent way. Interests, siRNA/CL-LA5 did not induce a exceedingly biocompatible intergalactic (IFN) reaction within the liver, as customary siRNAs did.

There are various vectors for gene therapy, including viral vectors such as retrovirus and adenovirus, as well as non-viral vectors such as liposomes (**Figure 3**). In gene therapy with liposomes, the first step is the construction of the liposome DNA complex. Cationic liposomes are routinely employed as one of the major nonviral transfecting agents for intracellular delivery of hydrophilic molecules such as nucleic acids, peptides, and proteins. Here, the purified DNA fragment is mixed with liposomes at room temperature in a serum-free environment. Then the liposome is injected intravenously to the patient. that gene therapy using liposomes is performed in the order of the following steps [19, 20]:

- Cationic liposomes are routinely employed as one of the major nonviral transfecting agents for intracellular delivery of hydrophilic molecules such as nucleic acids, peptides, and proteins. Cationic liposomes when complexed with DNA form a strong positively charged cationic liposome-DNA complex or lipoplex.

- Then liposomes are injected intravenously to the patient.

- Liposomes circulate in the blood to reach the target cell.

- The reaction of liposome with the target cell, which is usually the reaction of liposomes with the cell in two ways: endocytosis and fusion, which of course depends on their composition and method of preparation. Most cationic liposomes enter the cell through endocytosis [21].

- Release of DNA from the endosome, for this purpose amphipathic peptides are used in liposomes, a peptide with hydrophilic and hydrophobic domains at both ends of the alpha-helical structure, which is stable at pH = 7.4 but under acidic pH conditions, it is destroyed in the primary endosome, and the membrane becomes unstable, and the endosome ruptures [22].

DNA enters the nucleus through the nuclear pore complex. Small molecules up to 50 kDa or (310 bp) can enter the nucleus by passive diffusion, but larger molecules need to interact with importin α to enter the NLS signal and enter the cell by energy-dependent processes [23, 24].

6.2 The use of liposomes in immunotherapy

The era of clinically successful antitumor reactions regularly requires the effective execution of a few resistant forms (**Figure 1**). Firstly, various cancer antigens, either tumor-specific or tumor-associated antigens (TAAs), are discharged amid the method of tumor development. These cancer antigens are phagocytosed, prepared, and displayed by antigen-presenting cells (APCs) such as dendritic cells (DCs). At that point, the cancer antigens can be displayed into the major histocompatibility complex (MHC) course II particles or cross-presented into the MHC lesson I particles on DCs that relocate to depleting lymph hubs to start T cell actuation. Amid this prepare, DCs develop and costimulatory atoms (such as CD40, CD80, and CD86) are upregulated when particular prompts are display, such as damage-associated atomic design particles or pathogen-associated molecular pattern atoms display within the TME or given by implies of treatment. Upon maturation, DCs redesign their films to make dendrites to extend the layer surface range and improve T cell intuitive. In like manner, higher numbers of DCs show within the TME are useful and can make strides T cell enactmen.

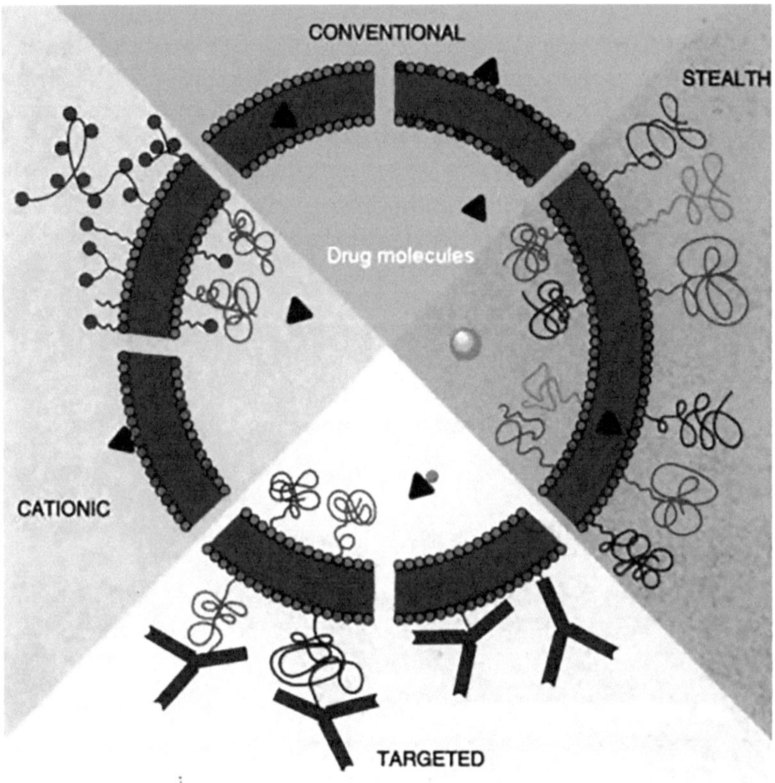

Figure 3.
Types of liposomes, normal liposomes, and safety of liposomes with high circulating power and cationic liposomes [3].

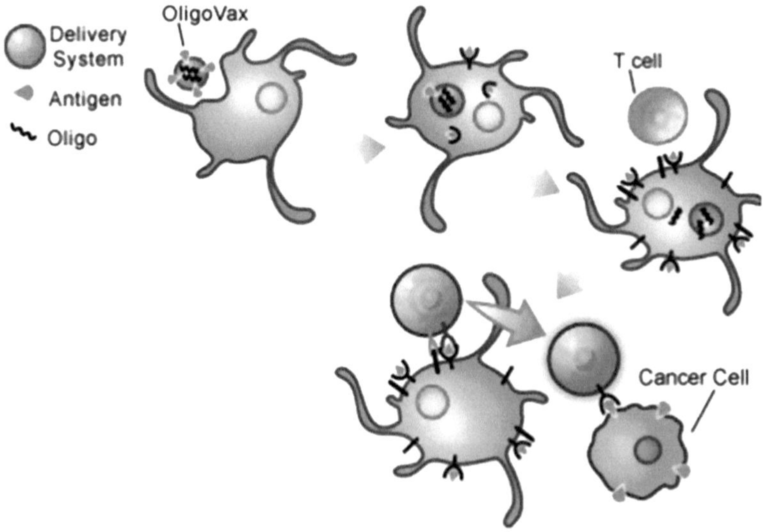

Figure 4.
Use of liposomes and placement of the Oligo vax vector inside it in order to stimulate the immune system against tumors and cancer cells [25].

Commercial companies have designed many vectors for this purpose, including Bioteach, which has designed a vector called Oligo vax (**Figure 4**), which is a liposomal vector containing antigens related to a disease on their surface along with a DNA molecule that stimulates the immune system. Antigens are taken from certain viruses, bacteria, or tumors. These antigens stimulate the production of immune system cells, and any cell that contains these antigens is destroyed. Also, short pieces of DNA are placed inside these liposomes, which are called oligos. Immune stimulatory sequences or ISS are among these sequences that are rich in guanine cytosine (GC) and stimulate the production of immune cells to a greater extent. In humans and other vertebrates, these sequences are rare, but they are abundant in bacteria that stimulate the immune system. Such liposomes are designed to stimulate immune cells against the tumor as much as possible [26].

6.3 The use of liposomes as drug carriers

One of the most important uses of liposomes is the release of drugs, especially in cases where there is a blood barrier such as the brain and spinal cord, this method of drug transfer and release is very efficient. Liposomes are used as drug carriers and for having amphipathic properties to deliver vaccines, enzymes, or drugs to the body. When liposomes are used as drug carriers, they protect healthy cells against the toxicity of drugs and prevent the concentration of drugs in vulnerable tissues such as the liver and kidney [27]. Using this method minimizes or completely eliminates the side effects of drugs. Today, liposomes are also used in cosmetics. Sometimes liposomes are used in topical skin treatments. This method is very efficient, because the liposome increases the penetration and absorption of the drug into the inner layers of the skin. Therefore, liposome is considered to be a very effective and efficient method of drug delivery for skin diseases, especially epidermal diseases. According to the nature of the drug, the drug can be embedded in the hydrophilic or hydrophobic layers of the liposome [28].

7. Methods of producing nanoliposomes

7.1 Thin layer hydration

The first production method was presented by a person named Bengham in 1965. It is a fast method to produce liposomes with high efficiency. Bengham's method, as a thin layer hydration method, is one of the most common methods of liposome formation. In this method, lecithin and the substances that are to be encapsulated are first dissolved in an organic solvent, then, the solvents are evaporated by an evaporator device, the dried lipid layer is hydrated in an aqueous environment such as a buffer, and the created suspension is stirred for a while until the hydration process is well done. Then, homogenization, sonication, and other size reduction methods are used to reduce the size and produce nanoliposomes [29, 30].

7.2 Detergent separation

This method is considered a suitable method for producing very homogeneous liposomes. The basis of this method is the formation of lipid-detergent micelles, and after that, liposomes are formed by separating the detergent. Lipid-detergent micelles are formed by hydrating the lipid with a detergent solution and then separating the detergent and organic solvent from the lipid and adding the aqueous solution.

The detergent comes in contact with the lipid and keeps the hydrophobic part from contacting the aqueous phase, and then, micelles are formed instead of vesicles. The detergent is separated from the micellar solution using methods such as dilution, dialysis, column chromatography, and absorption, and then liposomal vesicles are formed. One of the disadvantages of this method is that it is time-consuming, and during the separation of the detergent, there is a possibility of removing small hydrophilic compounds [31].

7.3 Ether and ethanol injection

In the ether and ethanol injection method, the lipid composition is first dissolved in an organic phase, and then, liposomes are formed by injecting the lipid solution into the aqueous medium. The ethanol injection method was first described by Batters by and Korn in 1973. In this method, a lipid solution dissolved in ethanol is injected into the aqueous environment using a thin needle, and this injection is done quickly. Since ethanol is injected directly into the water environment; Solvent retention has become a concern in this method, unless methods can be found to remove it after liposome formation. The ether injection method is completely different from the ethanol injection method due to its immiscibility in aqueous environment. In this method, the solvent can be removed from the liposomal product by using heat. In the ether injection method, unlike ethanol, the injection is done slowly [32].

7.4 Reverse-phase evaporation

The basis of this method is the formation of water droplets surrounded by lipid, which is dispersed inside the organic solvent. In this method, lipid compounds are first dissolved in an organic solvent, then the aqueous phase containing active substances is added to the organic phase, and after sonicating the solution, reverse micelles are formed. Finally, by removing the organic solvent using a solvent evaporation device, a viscous gel is formed. When a sufficient volume of solvent is removed, the gel breaks down, and an aqueous suspension of vesicles is formed [32].

8. The method of reducing the size of liposomes

Multilamellar vesicles can change after production, and this change can be in its size and degree of homogeneity. The most common methods used for size reduction include sonication, high pressure homogenization, microfluidizer, and colloidal mills.

8.1 Sonication

Sonication is a simple technique to reduce the size of liposomes, which ultimately leads to the production of nanoliposomes, especially SUV type liposomes. Two types of probes, namely sonicator and Bathroom sonicator, are used in the production of liposomes [33].

8.1.1 Probes sonicator

The hydrated vesicles are directly exposed to the sonicator probe, which has a sharp titanium tip, for several minutes so that the probe floats in the liposomal suspension. On the other hand, the probe is connected to a generator and by turning it on, and ultrasonic waves are emitted from the probe. Spreading these ultrasound waves with frequencies of several hundreds of kilohertz into the liposomal suspension causes the spontaneous production and disintegration of bubbles. The shock caused by the disintegration of the bubbles causes an increase in temperature and the creation of turbulent flows near the destroyed bubbles, and in this way, high shear forces are applied to the hydrated liposomes and lead to a decrease in their size. Since a lot of energy enters the solution through the probe, the solution heats up over time and leads to the collapse and instability of the liposomes; Therefore, to prevent this, it is necessary to use a suitable cooling system, which can be prevented from increasing the temperature by placing the solution in an ice Bathroom [33].

8.1.2 Bathroom sonicator

In the Bathroom sonicator, the solution is placed inside the tubes in the sonicator. The distribution of energy into the system is uniform, and since the energy transfer is indirect and does not come into contact with the solution, as a result, the system is less polluted and the liposomes are less destroyed. In this method, vesicles with uniform particle size distribution are produced. Due to being equipped with heating and cooling systems, temperature control is better, but it requires long times to reach the minimum particle size, and its effect is much less compared to the probe sonicator [30].

8.2 High pressure homogenization

In this method, the prepared liposomal dispersion is pumped into the chamber that has a head with a pressure of 10–100 MPa, which causes a strong pressure difference between the inner and outer parts of the head. When the solution passes through itself, the pressure drops so that the liquid boils and air bubbles form. When the solution stops boiling, the created bubbles burst after some time, and the shock caused by the bursting of the bubbles destroys and breaks the vesicles, resulting in the production of smaller vesicles. This phenomenon of homogenization is known as cavity theory of cavitation [31].

8.3 Micro fluidization

One of the methods of producing nanoliposomes without using toxic solvents is using a microfluidizer. In a microfluidizer, the liposomal premix is first introduced into a source and then pumped into the interacting chamber using air pressure. This chamber consists of two very narrow flow channels, thus creating two paths for the flow of liposomal suspension with large droplets. Finally, the liquid of these two paths collides at high speed, and in this way a very strong shear force is created, which causes the production of liposome with very small droplets. After the process, due to the use of high pressure and the rise in temperature of the product, a cooling system is used [31].

8.4 Colloid mills

Colloid mills are widely used in the food industry to homogenize liquids with medium and high viscosity. Also, this mill is used to reduce the particle size of liposomes. The colloidal mill has two disks: a rotor (rotating disk) and a stator (fixed disk). The rotating disc contains grooved surfaces to create turbulent flow. The solution is pumped into the device, and due to the speed difference between the rotor and stator surfaces, high shear forces are created, and this high shear force breaks the particles and creates smaller sized particles [31].

9. The use of nanoliposomes in the food industry

9.1 Application of liposome in dairy industry

One of the wide applications of liposomes in food products has been in the production of cheese. In the process of cheese production, adding protein before the formation of cheese curds reduces the time and cost of cheese ripening, but enzymes are often inactive under the conditions of the food environment. Adding unencapsulated free enzymes to milk causes premature proteolysis and thus causes the formation of a clot with an undesirable consistency. In this way, a large amount of enzymes is lost along with proteins, and due to the excessive use of enzymes in the beginning, the cost of producing the product also increases. By surrounding enzymes and creating a protective barrier around them, liposomes protect enzymes from adverse environmental factors. Also, enzymes can be added to the food before their activity is needed, without having the negative effect of adding free enzymes earlier. It has been reported that the addition of proteases in MLV liposomes to cheese prevented the rapid proteolysis of beta-casein, and as a result, the cheese texture was stable after fermentation and protected the clot structure from enzymatic attack, which is the same result for small lamellar ionic vesicles. It was obtained by the method of freezing and removing from freezing, which in this way caused the preservation of proteinases against adverse environmental factors. Other uses of liposome in the dairy industry include the production of low-lactose dairy products, and in this way, they can help people with lactose intolerance. For this purpose, the beta-galactoside enzyme is trapped inside the phosphatidylcholine-cholesterol vesicle, and in this way, this enzyme remains active even after a month of storage in the refrigerator and at pH = 3. Another use of liposomes, in the dairy industry, is to encapsulate preservatives. A clear example of this use; it is the preservation of cheeses whose curds are washed, such as: Gouda, Emmental, and Edam cheeses [32]. The environment of these types of cheeses is prone to the growth of various pathogenic microorganisms and spoilage. This type of spoilage can be controlled by adding nitrate to milk, but due to the ill effects of nitrate on health, there are many concerns regarding its use in food. Antimicrobial proteins such as lysozyme and enzymes derived from eggs are known as the best substitutes for nitrates in food preservation, but lysozyme is attached to milk casein, and thus its ability to inhibit spores is reduced. The liposome that surrounds the lysozyme prevents the formation of this connection and can also target the areas where this bacterium is present and destroy them by releasing the lysozyme inside it [33].

9.2 The use of liposomes in the release of antimicrobial substances

In recent years, many efforts have been made to access food additives and replace synthetic additives with organic and natural materials, but many of these natural materials are not as effective as synthetic materials or their high cost, has limited their use in food products. The use of encapsulation technology can overcome this weakness. So far, limited studies have been conducted in the field of antimicrobial substances encapsulated by liposomes as food additives. One of the first studies conducted in this field is the encapsulation of lysozyme or nisin as a protective factor for cheeses against microbial spoilage. Adding antibiotics directly to the cheese curd kills the starters, whose activity is of particular importance in the early stages of cheese production. When antimicrobial substances are encapsulated in liposomes, their release is controlled and delayed. In this way, the starters are protected from their destructive effects and can work well in the early stages of cheese production, but after being released, they can prevent the activity of undesirable bacteria. In 2009, the antimicrobial activity of thymol and carvacrol encapsulated in liposome was tested against four types of gram-positive bacteria and four types of gram-negative bacteria and food pathogens such as Listeria monocytogenes. All the tests indicated an increase in antimicrobial activity after applying the encapsulation technique [34, 35].

9.3 The use of liposomes in encapsulating antioxidants

Antioxidants work through different mechanisms and mainly inhibit free radicals. Antioxidants directly inhibit free radicals, reduce peroxide concentration, and repair oxidized membranes. They deactivate iron and reduce the products of active oxygen producing species. Short chain fatty acids neutralize reactive oxygen species. Antioxidants added to food are considered as nutritional compounds that have the ability to delay or prevent the development of rancidity and other flavor spoilages caused by oxidation. At the molecular and cellular level, antioxidants are able to deactivate free radicals. Nowadays, due to controversial issues, the use of natural antioxidants is preferred instead of synthetic ones in the food industry [25, 36].

10. Conclusion

- Due to the ease of transfer and the absence of obstacles in the way of transfer of liposomes, the use of liposomes has been investigated as a way to transfer drugs and treat some diseases, and in fact, in this way, some toxic drugs that are anti-tumor or to transfer drugs that are not easily absorbed in the body to the target tissue. So that with the application of this biological system, many obstacles on the way of drug delivery to different tissues will be removed, and a new chapter will be opened for the development and progress of new fields in medicine, including nanomedicine. In fact, the purpose of using this biological system is that by injecting liposomes into the circulatory system, they are selected and absorbed by tumor cells, and after they are broken at the site of tumor cells, they are sprayed on cancer cells and drugs are administered without. They cause damage to the healthy cells of the body. It should be noted that, in the use of liposomes for gene transfer, the preparation and packaging of foreign DNA is easily done, and with this method, large pieces of DNA can be transferred. Compared to viral vectors, these vectors are non-toxic and do not stimulate the immune system. They can also be prepared in large quantities and easily manipulated. In order to use liposomes as drug carriers, liposomes can easily cross blood barriers and release drugs in the brain. Despite the multiplicity of advantages and applications of liposomes, this biological system, like other biological systems, has some disadvantages, including the low transfer efficiency and the limited durability and stability of liposomes, and also in this method, the gene cannot be embedded in the genome and gene expression. The target is low. So that in order to use liposomes in clinical applications, many structural, functional, physical, and chemical aspects of liposomes should be considered, so that the application of this biological system requires more investigations.

- Due to their biocompatibility, liposomes and nanoliposomes have many applications in the release and delivery of drugs and cosmetics, and their introduction into the food industry has brought about a significant change in it. Liposomes have many uses in the food industry, including the preservation of sensitive food components, and in this way, the effectiveness of food additives can be increased. With surrounding compounds and creating a coating around them, liposomes protect food components against environmental and chemical changes and increase their stability against extreme changes in temperature and pH. The study of recent research indicates that liposome has made significant progress in the dairy industry, including cheese. Among other progressing researches of liposomes in the food industry is the encapsulation of antimicrobial compounds to protect food materials from spoilage, and in this way, the quality and health of food components can be increased in a wide range of food compounds.

Author details

Ehsan Kianfar[1*], Abolfazl Mirani[1], Zaid H. Mahmoud[2] and Arezoo Mohammadkhani[3]

1 Biomedical Engineering Research Center, Clinical Sciences Institute, Baqiyatallah University of Medical Sciences, Tehran, Iran

2 Department of Chemistry, College of Sciences, University of Diyala, Iraq

3 Department of Chemistry, Faculty of Science, Firooz Abad Branch, Islamic Azad University, Fars, Iran

*Address all correspondence to: ehsan_kianfar2010@yahoo.com; ehsankianfar775@gmail.com

References

[1] Barba AA, Bochicchio S, Dalmoro A, Lamberti G. Lipid delivery systems for nucleic-acid-based-drugs: From production to clinical applications. Pharmaceutics. 2019;**11**(8):360

[2] Olusanya TO, Haj Ahmad RR, Ibegbu DM, Smith JR, Elkordy AA. Liposomal drug delivery systems and anticancer drugs. Molecules. 2018;**23**(4):907

[3] Gabizon AA. Liposomal drug carrier systems in cancer chemotherapy: Current status and future prospects. Journal of Drug Targeting. 2002;**10**(7):535-538

[4] Chatterjee S, Banerjee DK. Preparation, isolation, and characterization of liposomes containing natural and synthetic lipids. In: Basu SC, Basu M, editors. Liposome Methods and Protocols. Methods in Molecular Biology. Vol. 199. Humana Press. 2002. pp. 3-16. DOI: 10.1385/1-59259-175-2:03

[5] Gardlik R, Pálffy R, Hodosy J, Lukács J, Turna J, Celec P. Vectors and delivery systems in gene therapy. Medical Science Monitor. 2005;**11**(4):110-121

[6] Lee TW, Matthews DA, Blair GE. Novel molecular approaches to cystic fibrosis gene therapy. Biochemical Journal. 2005;**387**(1):1-15

[7] Duncan R. Nanomedicine gets clinical. Materials Today. 2005;**8**(8):16-17

[8] Freitas RA Jr. Nanotechnology, nanomedicine and nanosurgery. International Journal of Surgery. 2005;**3**(4):243-246

[9] Hashida M, Kawakami S, Yamashita F. Lipid carrier systems for targeted drug and gene delivery. Chemical and Pharmaceutical Bulletin. 2005;**53**(8):871-880

[10] Worgall S. A realistic chance for gene therapy in the near future. Pediatric Nephrology. 2005;**20**:118-124

[11] Laura Immordino M, Dosio F, Cattel L. Stealth liposomes: Review of the basic science, rationale, and clinical applications, existing and potential. International Journal of Nanomedicine. 2006;**1**(3):297-315

[12] Kay, Son K. Cationic liposome gene transfer. In: Walther W, Stein U, editors. Gene Therapy of Cancer: Methods in Molecular Medicine™. Vol. 35. Humana Press. 2000. pp. 323-329. DOI: 10.1385/1-59259-086-1:323

[13] Barenholz Y. Liposome application: Problems and prospects. Current Opinion in Colloid & Interface Science. 2001;**6**(1):66-77

[14] Bally MB, Zhang YP, Wong FM, Kong S, Wasan E, Reimer DL. Lipid/DNA complexes as an intermediate in the preparation of particles for gene transfer: An alternative to cationic liposome/DNA aggregates. Advanced Drug Delivery Reviews. 1997;**24**(2-3):275-290

[15] Tseng WC, Huang L. Liposome-based gene therapy. Pharmaceutical Science & Technology Today. 1998;**1**(5):206-213

[16] Schmid MH, Korting HC. Therapeutic progress with topical liposome drugs for skin disease. Advanced Drug Delivery Reviews. 1996;**18**(3):335-342

[17] Kaur IP, Kapila M, Agrawal R. Role of novel delivery systems in developing topical antioxidants as therapeutics to

combat photoageing. Ageing Research Reviews. 2007;**6**(4):271-288

[18] Gouin S. Microencapsulation: Industrial appraisal of existing technologies and trends. Trends in Food Science & Technology. 2004;**15**(7-8):330-347

[19] Sharma Vijay K, Mishra D, Sharma A, Srivastava B. Liposomes: Present prospective and future challenges. International Journal of Current Pharmaceutical Review & Research. 2010;**1**(2):6-16

[20] Ohvo-Rekilä H, Ramstedt B, Leppimäki P, Slotte JP. Cholesterol interactions with phospholipids in membranes. Progress in Lipid Research. 2002;**41**(1):66-97

[21] Liang X, Mao G, Ng KS. Mechanical properties and stability measurement of cholesterol-containing liposome on mica by atomic force microscopy. Journal of Colloid and Interface Science. 2004;**278**(1):53-62

[22] Bangham AD, Standish MM, Watkins JC. Diffusion of univalent ions across the lamellae of swollen phospholipids. Journal of Molecular Biology. 1965;**13**(1):238-IN27

[23] Meure LA, Foster NR, Dehghani F. Conventional and dense gas techniques for the production of liposomes: A review. AAPS PharmSciTech. 2008;**9**:798-809

[24] Panwar P, Pandey B, Lakhera PC, Singh KP. Preparation, characterization, and in vitro release study of albendazole-encapsulated nanosize liposomes. International Journal of Nanomedicine. 2010;**5**:101-108. DOI: 10.2147/IJN.S8030

[25] Kianfar E. Protein nanoparticles in drug delivery: Animal protein, plant proteins and protein cages, albumin nanoparticles. Journal of Nanobiotechnology. 2021;**19**(1):159

[26] Mozafari M, Nanoliposomes R. Preparation and analysis. In: Liposomes: Methods and Protocols, Volume 1: Pharmaceutical Nanocarriers. 2010. pp. 29-50

[27] Richardson ES, Pitt WG, Woodbury DJ. The role of cavitation in liposome formation. Biophysical Journal. 2007;**93**(12):4100-4107

[28] Akbarzadeh A, Rezaei-Sadabady R, Davaran S, Joo SW, Zarghami N, Hanifehpour Y, et al. Liposome: Classification, preparation, and applications. Nanoscale Research Letters. 2013;**8**:1-9

[29] Pupo E, Padrón A, Santana E, Sotolongo J, Quintana D, Dueñas S, et al. Preparation of plasmid DNA-containing liposomes using a high-pressure homogenization–Extrusion technique. Journal of Controlled Release. 2005;**104**(2):379-396

[30] Kirby CJ, Brooker BE, Law BA. Accelerated ripening of cheese using liposome-encapsulated enzyme. International Journal of Food Science & Technology. 1987;**22**(4):355-375

[31] Reza Mozafari M, Johnson C, Hatziantoniou S, Demetzos C. Nanoliposomes and their applications in food nanotechnology. Journal of Liposome Research. 2008;**18**(4):309-327

[32] Liolios CC, Gortzi O, Lalas S, Tsaknis J, Chinou I. Liposomal incorporation of carvacrol and thymol isolated from the essential oil of Origanum dictamnus L. and in vitro antimicrobial activity. Food Chemistry. 2009;**112**(1):77-83

[33] Maherani B, Arab-Tehrany E, Kheirolomoom A, Cleymand F, Linder M. Influence of lipid composition on physicochemical properties of nanoliposomes encapsulating natural dipeptide antioxidant l-carnosine. Food Chemistry. 2012;**134**(2):632-640

[34] Mahmoud ZH, Ajaj Y, Hussein AM, Al-Salman HNK, Mustafa MA, Kadhum EH, et al. CdIn2Se4@ chitosan heterojunction nanocomposite with ultrahigh photocatalytic activity under sunlight driven photodegradation of organic pollutants. International Journal of Biological Macromolecules. 2024:131465

[35] Hsu CY, Rheima AM, Kadhim MM, Ahmed NN, Mohammed SH, Abbas FH, et al. An overview of nanoparticles in drug delivery: Properties and applications. South African Journal of Chemical Engineering. 2023;**46**:233-270

[36] Kianfar E. Magnetic nanoparticles in targeted drug delivery: A review. Journal of Superconductivity and Novel Magnetism. 2021;**34**(7):1709-1735

Section 3

Liposomes for Pharmaceutical Delivery

Chapter 4

Revolutionizing Therapy: Nanomaterials in Liposomes Redefine the Future of Medicinal Drugs

Jerusa M. Oliveira, Rener M.F. Duarte, Samaysa de L. Lins, Lívia M.S. de Lima, Jéssica M. Pereira, Larissa I.M. de Almeida, Dhandara E. de L. Sampaio, Auana R. da S. Andrade, Isabella de O.F. de Sousa, Carlo J.F. Oliveira, Virmondes Rodrigues, Marcos V. da Silva, Foued Salmen Espindola, Fabiane C. de Abreu, Lucas Anhezini, Juliana Reis Machado e Silva and Anielle Christine A. Silva

Abstract

Liposomes are microscopic lipid-based vesicles that have emerged as a promising vehicle for transporting therapeutic agents with precision and efficiency. From enhanced drug bioavailability to targeted delivery, combining nanomaterials and liposomes offers a transformative approach to therapeutic interventions. Encapsulating nanomaterials with drugs in liposomes holds immense significance as it enhances precision, efficiency, and targeted delivery, revolutionizing therapeutic interventions in medicine. This chapter delves into the unique properties of nanomaterials encapsulated within liposomes, examining their potential to revolutionize medicine. In addition, it highlights key advancements, challenges, and prospects in this dynamic and rapidly evolving field, providing readers with a comprehensive understanding of the revolutionary impact on the future of medicinal drugs.

Keywords: nanomaterials, drugs, plant extracts, biocompatibility, therapeutic agents, co-encapsulating

1. Introduction

The efficient delivery of medications is a constant challenge in the healthcare field, and liposomes (LPs) emerge as a promising solution to overcome many of the limitations associated with conventional drug administration. These bilayered lipid vesicles have stood out due to their remarkable advantages in terms of bioavailability, controlled drug release, and specific targeting. The ability to encapsulate both hydrophilic and lipophilic drugs confers unique versatility to liposomes, expanding the spectrum of therapeutic substances that can be effectively administered [1–3].

The lipid structure of liposomes also provides a valuable protective layer, beneficial for drugs sensitive to enzymatic degradation or other external agents [4]. Controlled release and the ability to target drug delivery to specific tissues contribute to reducing side effects and achieving more efficient administration. The capacity to combine different therapies within a single liposomal system offers a multifaceted approach to treating various medical conditions [5–7]. Combining other therapies within a single liposomal system provides a multifaceted and highly versatile approach to treating various medical conditions. This strategy enables the coordinated delivery of multiple drugs, gene therapies, or contrast agents, allowing for a more comprehensive and effective approach to managing complex diseases. This therapeutic synergy can significantly improve clinical outcomes, providing new opportunities for personalized treatments tailored to patients' needs.

Liposomes also have demonstrated usefulness in imaging diagnostic applications, acting as contrast agents that enhance the visualization of specific body areas. Their low toxicity and ability to improve the stability of unstable drugs further extend their clinical potential, offering a safe and effective approach to enhance diagnostic accuracy and therapeutic monitoring in various medical conditions [8–10].

Collectively, these characteristics make liposomes a remarkable tool in drug delivery, promoting significant advancements in therapeutic efficacy and the minimization of adverse effects (**Figure 1**). Thus, liposomes can contribute to the ongoing progress of modern medicine. This chapter aims to demonstrate the importance and advancement of liposome utilization in biomedicine. We compiled works published in

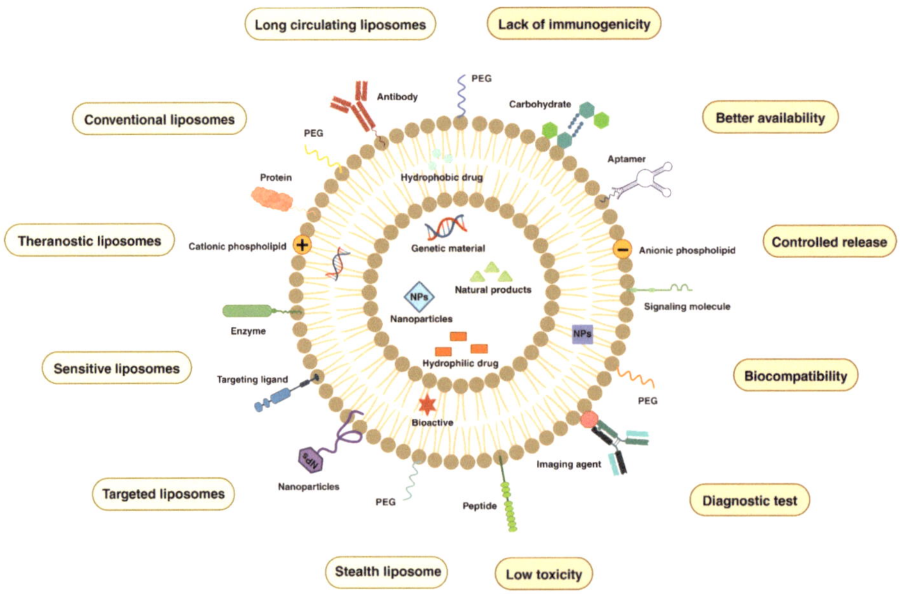

Figure 1.
Schematic representation of different types of liposomes for drug delivery and other bioactives. In the image, we observe various classifications of liposomes, highlighted by the presence of polyethylene glycol (PEG), electrostatic charges, antibodies, enzymes, peptides, carbohydrates, imaging agents, and other specific ligands. Due to their numerous functionalities, their advantages, safety, and toxicity are highlighted in the figure, emphasizing their promising use in nanomedicine.

the last five years from three databases (Web of Science, PubMed, and Virtual Health Library) to achieve this. We described their main contributions and perspectives for scientific advancement. Our search was conducted using the following descriptors: "Liposome," "Pharmaceutical Preparations," and "Nanoparticles." Accordingly, we illustrate the progress and advantages of incorporating nanomaterials, drugs, and natural products used in treating various diseases into liposomes and the various functionalizations of liposomes (**Figure 1**).

2. Incorporation of nanomaterials into liposomes

Utilizing inorganic nanoparticles in liposomes represents an innovative strategy for developing drug delivery systems. This approach aims to enhance liposomes' physical, chemical, and therapeutic properties, endowing them with specific characteristics crucial for advanced clinical applications. Such formulations can also be employed for loading liposomes with luminescent nanomaterials, such as rare earth elements, quantum dots (QDs), or silica, thereby creating bioimages for theranostics.

Theranostic materials with dual diagnostic and therapeutic properties facilitate the localization and/or tracking of liposomes within the organism or in cultured cells while enhancing contrast in imaging modalities such as magnetic resonance imaging (MRI) and X-ray imaging, among others [11–13]. To study the physiology of metabolic diseases such as diabetes and obesity [14], the liposomes with lanthanide nanoparticles (NPs@Lips) enable imaging acquisition through the near-infrared I window (NIR-I, 700–900 nm). Following *in vitro* and *in vivo* testing, the authors successfully visualized the inter-scapular brown adipose tissue (BAT) and distinguished it from the white adipose tissue (WAT) in NIR-I bioimages using the NPs@Lips with low toxicity. Similarly, in a parallel study, when Ytterbium (Yb^{3+}) was incorporated into the liposome formulation with doxorubicin, the element was sensitized, generating emission in the near-infrared (NIR) region, thereby enabling the monitoring of drug release [15]. Thus, using rare earth NPs and a lanthanide can add potential for multifunctional applications in nanomedicine.

Various types of inorganic nanoparticles have been incorporated into lipid vesicles, each playing a unique role for specific purposes [5, 16, 17]. Some examples are shown in **Table 1**. The incorporation of nanoparticles (NPs) into liposomes (LipoNPs) such as metallic gold (Au), zinc oxide (ZnO), and silver (Ag) NPs aims to improve the stability of liposomes and provide controlled drug release. Additionally, adding carbon nanotubes and modified polymers strengthens the mechanical properties and extends the half-life of liposomes. Meanwhile, the presence of magnetic nanoparticles offers the ability to target and control the delivery of these vesicles through external magnetic fields. At the same time, modified polymeric dendrimers alter the surface of liposomes to enhance drug transport capabilities [7, 17, 18].

These strategies offer remarkable benefits, such as improved stability, controlled drug release, increased loading capacity, directional guidance, optical properties for imaging diagnostics, and the potential for therapy combination. The combination of different nanoparticle-mediated therapies may lead to a synergistic therapeutic outcome, enhancing various types of cancer treatments [17, 18], infectious diseases, fungal infections, antibacterial therapies [19], and medical imaging [20]. The careful selection of inorganic nanoparticles to be incorporated into liposomes depends on the intended clinical application, highlighting the versatility of this innovative approach in the field of drug delivery.

Nanoparticles (NPs)	Liposomes type	Applications	Nanostructures' size	Author
Iron oxide	soy phospholipids and cholesterol	Treatment for anemia	125 ± 5.8 nm	[23]
Cooper	soy lecithin	Breast cancer treatment	≥ 100 nm	[30]
Gold	Conventional liposomes	in vitro: cancer cell lines of osteosarcoma (U2OS)	131.1 ± 20.1 nm	[49]
Ytterbium	Conventional liposomes	*in vivo*: mouse	~50 nm	[15]
CdSe/CdSe	Conventional liposomes	diagnosis bioimage	2–5 nm	[34]
Lanthanide	Conventional liposomes	in vitro: human liver cells in vivo: mouse	71.4 ± 1.5 nm	[14]
Platinum	Biotin-modified liposomes	Clinical diagnosis and treatment of diseases	100 nm	[35]
Iron oxide coated with citric acid	Thermosensitive liposomes	in vitro: breast cancer cells	8.11 ± 1.12 nm	[24]
Iron oxide coated with citric acid	Cationic liposomes	in vitro: U87 human primary glioblastoma cells	< 20 nm	[25]
Gold NPs	Cationic liposomes	in vitro: human colorectal cancer cell *in vivo*: mouse	100 nm - 140 nm	[31]

CdSe/CdSe = CdSe/CdS_xSe_{1-x} *magic-sized quantum dots (MSQDs).*

Table 1.
Incorporations and applications of the nanoparticles into liposomes in biomedicine.

Anemia is a public health issue prevalent worldwide and affects individuals of all ages, with women, pregnant women, children, and the elderly being particularly vulnerable. This condition can have serious health consequences, including fatigue, weakness, difficulty concentrating, impaired cognitive development in children, pregnancy complications, and untreated cases that may lead to death [21, 22]. The primary treatment for anemia involves the intake of iron pills, which may cause discomfort for the patient. Considering this, Fathy et al. [23] encapsulated magnetic iron oxide nanoparticles (MNPs) within liposomes, creating liposomes loaded with MNPs (LMNPs) to investigate their efficacy in treating iron-deficiency anemia. LMNPs were engineered to enhance the stability and surface properties of MNPs, allowing them to evade the reticuloendothelial system (RES) and avoid opsonization. A study on female rats revealed that oral administration of LMNPs for 13 days was more effective in treating anemia. They successfully restored hematological parameters from anemic to normal levels. This outcome holds promise for advancing the efficiency of anemia treatment. Furthermore, MNPs encapsulated in thermosensitive liposomes (TSLs) could present a promising therapeutic strategy for enhanced treatment of breast cancer [24, 25].

Numerous studies highlight cancer as one of the leading causes of mortality worldwide. However, treating this disease poses notable limitations, especially in systemic chemotherapy, where low drug concentration in the tumor, coupled with rapid elimination from circulation, results in significant toxic effects outside the tumor region [26]. Liposome and NP complexes are also being tested and utilized to diagnose and treat different cancer types and effectively transport anticancer drugs. These liposome complexes have the property of maximizing therapeutic efficacy and minimizing side effects of pure metallic complexes that exhibit efficiency in treating the disease, such as Ag, Au, copper (Cu), and nickel (Ni) NPs [27–29]. For instance, Cu NPs, when encapsulated in liposomes synthesized from soy lecithin, showed promising and effective results in breast cancer treatment and were less toxic to normal cells. The Cu NPs in liposomes demonstrated improved safety profiles in MCF-7 cells compared to their free form [30].

Gold nanoparticles (GNPs) and carboplatin encapsulated in liposomes (LipoGold) were tested against human colorectal carcinoma cells (HCT-116). The LipoGold was produced using the scaled-up microfluidic fishbone method and administered simultaneously. This study [31] demonstrates that drug and gold nanoparticles (AuNPs) encapsulation enhances the efficiency in reducing cancer cell proliferation. LipoGold significantly delays tumor growth compared to other formulations, even at lower doses. Additionally, it was possible to suggest that the LipoGold formulation enables a more realistic *in vivo* treatment with significantly lower amounts of GNPs, which may allow for greater clinical viability.

Carbon nanotubes (CNTs) are hollow graphitic nanomaterials with diameters ranging from 2 to 20 nm and extremely high aspect ratios. This nanomaterial can also be utilized as a novel drug delivery mechanism. They can be loaded with drugs and covalently attached to CNTs to form CLC (Carbon Nanotube Liposome Complex). This approach combines the efficient cellular uptake of CNTs with liposome's well-known high drug-loading capacity. This combination can significantly reduce the toxicity associated with free CNTs and offer the potential for more effective and targeted combination therapies [32], such as cancer treatment [33].

Innovative drug delivery systems utilizing liposomes and nanocrystals also stand out in their potential applications in theranostics (therapy and diagnosis) [34]. Liposomes containing CdSe/CdS magic-sized quantum dots (MSQDs) have the potential to serve as stable fluorescent reporters and, consequently, can be explored as an innovative luminescent tool for drug delivery. The co-encapsulation of MSQDs within liposomes enables precise monitoring of their spatial distribution through luminescent emission while facilitating targeted delivery of drugs to desired sites [34].

The application of liposomes in single-particle collision electrochemical biosensors (SPCE) for detecting H9N2 avian influenza virus (H9N2 AIV) is promising. This biosensor is constructed by integrating liposome release strategy with immunomagnetic separation. Liposomes, modified with biotin and loaded with platinum nanoparticles (Pt NPs), act as signal probes for virus detection. Upon contact with H9N2 AIV, controlled release of Pt NPs from liposomes occurs, resulting in a detectable increase in electrochemical signal. This strategy enables enhanced electrochemical sensitivity for precise and rapid detection of H9N2 AIV in clinical samples. Liposomes' controllable release amplification strategy offers significant versatility for applications in other biosensors and detection systems. This combination of nanotechnologies holds great potential in contributing to developing new technologies for rapid and accurate diagnosis of viral diseases [35].

The encapsulation of nanoparticles (NPs) within liposomes offers an innovative and promising approach to developing advanced drug delivery systems. Combining different types of NPs within liposomes, yields benefits in enhancing stability, controlling drug release, increasing loading capacity, providing directional targeting, and reducing toxicity. Furthermore, liposome-nanoparticle complexes have shown efficacy in cancer treatment while mitigating the side effects associated with conventional therapies. However, studies still have significant gaps regarding incorporating NPs into liposomes. One such gap is the long-term stability of liposomes, which may impact their therapeutic efficacy. Additionally, further investigations are required on the toxicity of formulations to normal cells, immune responses in various cell types and *in vivo*, and better standardization of synthesis methods. Given the promise of these formulations, LipoNPs could be further investigated and explored for application in the treatment of other diseases, such as neglected diseases, and provide more efficient and targeted drug delivery with reduced side effects for various clinical applications.

3. Nanoparticle-liposome hybrid systems: revolutionizing pharmaceuticals in nanomedicine

In recent years, biomedical research has been intensively dedicated to developing innovative strategies for efficient drug delivery. Biotechnology aims to enhance therapeutic efficacy, reduce potential adverse effects such as side effects and toxicity, and address issues such as low solubility. A prominent approach in this scenario is incorporating drugs, natural products, or their bioactive substances into liposomes. This versatile technique has been applied to various drugs, ranging from chemotherapeutic agents, nanoparticles, and nanoparticles adsorbed onto natural products such as plant extracts and essential oils to pharmaceuticals and vaccines, standing out as a promising platform in clinical applications [36–39].

Liposomes enable the encapsulation of both water-soluble and lipid-soluble compounds. For hydrophilic substances like many conventional chemotherapeutic agents, liposomes provide an aqueous environment within their interior, facilitating solubilization and stability of these substances during administration. For lipophilic drugs, the lipid bilayer of liposomes provides a hydrophobic compartment, allowing for the incorporation and bioavailability of these substances.

Moxifloxacin hydrochloride (MOX) encapsulated in liposomes tested against *Staphylococcus epidermidis* demonstrated greater efficacy in inhibiting bacterial growth and reducing the formation of bacterial biofilms. In terms of public health, effective control of biofilm growth by *S. epidermidis* and other bacteria associated with medical devices is crucial for reducing rates of hospital-acquired infections and improving the safety of medical procedures [40]. Incorporating antibiotics into liposomes, such as Tylosin, is also a promising strategy for effectively treating antibiotic-resistant bacterial pathogens [41].

Simvastatin (STAT) is clinically prescribed orally to reduce serum cholesterol and can be used as an anti-inflammatory. However, STAT has low bioavailability in the bloodstream and needs to be administered at high doses, which can cause undesired effects. Nevertheless, when STAT was encapsulated in liposomes, the drug could be delivered directly to the atherosclerotic plaque for anti-atherosclerotic benefit. The formulation's administration reduced inflammation and increased cholesterol efflux in 2D and 3D cell models. Additionally, it reduced the secretion of pro-inflammatory cytokines and the expression of cell adhesion molecules, suggesting a potential anti-inflammatory and lipid-reducing effect [42].

In oncology, the ability of liposomes to selectively accumulate in tumors, leveraging the phenomenon known as the "enhanced permeability and retention" (EPR) effect, has allowed for a more targeted delivery of antitumor agents, thereby reducing side effects in healthy tissues [43, 44]. An example is the incorporation of tuftsin co-encapsulated with doxorubicin and curcumin in liposomes for cancer treatment. Tuftsin is a tetrapeptide that enhances the anti-tumorigenic potential of drugs encapsulated in liposomes. The innovative formulation inhibited Ehrlich ascites carcinoma tumor growth in rats and human cervical cancer cell lines (HeLa). Adding curcumin can act as a chemosensitizer to reverse doxorubicin resistance against solid tumors [45].

Liposomes can also overcome barriers to using plant extracts for disease treatment [38]. Loading liposomes with natural products (e.g., plant extracts, essential oils, propolis, bioactive compounds, and natural antibiotics) surpasses existing limitations. It can also make versatile formulations for treating cancer and other diseases [37, 46, 47]. For example, Melchior et al. [48] synthesized and characterized liposomes loaded with quercetin. The authors observed that encapsulation of the natural product was more efficient in reducing the viability of colon cancer cells (HCT-116 p53+/+ cells) compared to free quercetin. Similarly, in the study by Melchior et al. [48], the synthesis and characterization of liposomes loaded with quercetin demonstrated promising results. The researchers noted that encapsulation of this natural compound resulted in a more efficient reduction of HCT-116 p53+/+ cell viability than free quercetin. These findings underscore the potential of liposomes as delivery vehicles to enhance the therapeutic efficacy of bioactive compounds, offering new perspectives for treating cancer and other pathologies.

As previously mentioned, the conjugation of nanomaterials with liposomes has been employed to enhance the stability of formulations, which aids in the specific targeting of drugs. This enhancement significantly contributes to preserving the therapeutic efficacy of encapsulated medications. Leveraging this versatility, creating

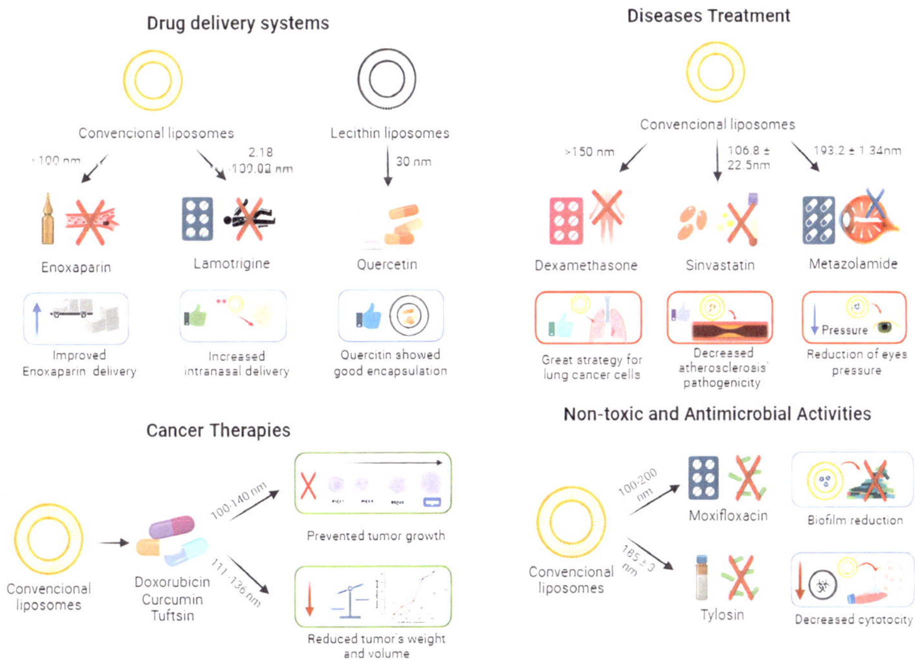

Figure 2.
Nanoparticle-liposome hybrid systems encapsulated with active plant or pharmaceutical principles for treating different human pathologies. References citations [40–42, 45, 48, 63].

formulations combining nanoparticles, natural products, and liposomes is important for nanotechnology. This formulation improves the delivery of natural products, provides more protection and stability for extracts, and can enhance synergistic effects. Thus, creating formulations combining nanoparticles, natural products, and liposomes plays a significant role in nanotechnology. These formulations improve the delivery of natural products and offer greater protection and stability for the extracts, enhancing synergistic effects and opening new perspectives for developing more effective and safer therapies.

Additionally, they can potentiate therapeutic benefits, reduce the likelihood of drug resistance, increase bioavailability, and mitigate toxicity and associated side effects of treatment [36]. However, more studies still need to utilize this type of formulation. Our group has been developing formulations containing different natural products, such as curcumin and quercetin, adsorbed onto inorganic nanoparticles. These formulations have shown low toxicity and promising results in treating neurodegenerative diseases in *in vivo* experiments (unpublished data).

Therefore, co-encapsulating liposomes will significantly expand the spectrum of therapeutic agents that can be effectively delivered and broaden the possibilities for therapies in nanomedicine. Combining nanomaterials with liposomes represents an innovative approach to drug delivery, opening doors to significant advancements in therapeutic efficacy and minimizing undesired side effects. This technological convergence may redefine paradigms in medicine and offer adaptable and efficient solutions for a diverse range of applications in nanobiomedicine (**Figure 2**).

4. Functionalization of liposomes

The functionalization of liposomes, both internally and externally, has emerged as an innovative and versatile strategy in drug delivery, providing significant advancements in optimizing therapeutic efficacy. This biotechnological phenomenon refers to the controlled and personalized modification of the surface or interior of liposomes to confer specific properties that go beyond the natural characteristics of these lipid vesicles. Both internal and external functionalizations of liposomes have their distinct methodologies, advantages, and disadvantages, delineating a fascinating research field with profound implications for various clinical applications.

Internal functionalization of liposomes involves incorporating bioactive molecules or nanomaterials within these vesicles' lipid core or lipid bilayer. Frequently used methods include co-lipophilization, where hydrophobic compounds are encapsulated along with lipids during liposome formation. This process is particularly effective for protecting sensitive drugs and improving stability, allowing for controlled and targeted release. However, challenges associated with encapsulation efficiency and potential adverse interactions between internal components require careful consideration during internal functionalization.

External functionalization, on the other hand, focuses on modifying the surface of liposomes, allowing for the conjugation of specific ligands, polymers, or nanoparticles to the outer layer. Methodologies include these entities' covalent or non-covalent attachment to the liposomal surface, providing precise customization of surface properties. This approach is valuable for selectively targeting liposomes to specific tissues or cells, enhancing the selectivity of drug delivery and diagnostics. For example, the formulation of intra-liposomal GNPs functionalized with SPE-PEG (2000)-maleimide-peptide demonstrated efficiency in penetrating mitochondria and inducing biological autoluminescence, directly impacting the functioning of these organelles [49]. Nanocarriers like this play a fundamental role in biological research, significantly contributing to understanding biochemical and physiological processes. Moreover, they have vast potential applications, including use in biosensor devices and detection. In nanomedicine, their versatility is evident, being employed for imaging live cells, real-time monitoring of biological processes, and early disease detection. This wide range of applications underscores the importance of nanocarriers in scientific research and the advancement of medicine.

In terms of applications, the functionalization of liposomes has gained prominence in domains such as oncologic therapy, where targeted delivery can maximize the effectiveness of antitumor agents. Additionally, in regenerative medicine and gene therapy, the functionalization of liposomes has shown promising implications for the efficient delivery of nucleic acids and growth factors. The highlighted applications of liposome functionalization extend to areas like oncologic therapy, where targeted delivery can enhance the efficacy of antitumor agents. Moreover, in regenerative medicine and gene therapy, liposome functionalization has demonstrated promising implications for the efficient delivery of nucleic acids and growth factors, paving the way for more effective and personalized treatments.

The functionalization of liposomes can also be utilized to achieve targeted drug delivery while circumventing the individual's immune system. Due to their biocompatibility, phagocytes readily capture liposomes, rapidly removing them from the bloodstream. Stealth liposomes, which feature a coating of inert and biocompatible polymers, have been developed to overcome this scenario. Poly(ethylene glycol) (PEG) is the most widely employed polymer for this purpose, thereby preventing the drug or biomolecule transported by the liposome from triggering an immune response in the body. This strategy aims to prolong the circulation time of liposomes and improve the efficacy of therapeutic loading, minimizing detection and removal by the phagocytic system [50].

Fusogenic liposomes can be considered an ideal tool for adoptive cell therapy (ACT). This liposome can fuse with biological membranes, thereby increasing contact of the medication and delivery into cells. Zheng et al. [51] developed a fusogenic liposome termed anti-phagocytosis-blocking repolarization-resistant membrane-fusogenic liposome (ARMFUL). This liposome features a core-shell structure, with a CSF1 receptor inhibitor (BLZ945) and anti-CD47 (aCD47) conjugated to the fusogenic lipid surface. ARMFUL was fused with the cell membrane of M1 macrophages, ensuring effective phagocytosis of tumor cells under CD47 antiphagocytic blockade. Additionally, ARMFUL/M1 effectively inhibited the growth of the mouse melanoma cell line (B16F10) and activated T cell-mediated immunity to suppress distant tumors, preventing tumor metastasis.

Therefore, ARMFUL proved to be a versatile tool for the synchronized engineering of adoptive cells, offering a platform for multimodal customization of cellular functions and behaviors. This promotes improvements in ACT against tumors, indicating significant potential for advanced therapeutic applications.

In biomimetic nanotechnology using RNAi, hybrid nanovesicles are developed. This type of nanovesicle was tested by ref. [52] in the treatment against non-small cell lung carcinoma (NSCLC), which accounts for approximately 85% of lung neoplasms [53]. The authors constructed the nanovesicles with cancer cell membranes (Cm) and charge-reversal liposome membranes (Lipm), using switchable matrix metallopeptidase 9 (MMP-9) peptides to coat polypeptides modified with lipoic acid (LC). These polypeptides are co-loaded with phosphoglycerate mutase 1 (PGAM1) siRNA and docetaxel (DTX). The coating in the intermediate layer was negatively charged (poly-L-lysine grafted with citraconic anhydride, PC), enabling pH-triggered charge conversion functionalization. Research results demonstrated that the integrated hybrid nanovesicle exhibits prolonged circulation half-life, effective lung cancer targeting, biocompatibility, high tumor accumulation, penetration into MMP-9 activated tumor cells, pH and redox-triggered DTX, and siRNA release.

Developed to co-deliver siPGAM1 and DTX, the nanovesicle showed a synergistic effect on tumor inhibition *in vitro* and *in vivo*, regulating glycolysis without notable toxicity and prolonging the lifespan of xenografted mice. The innovative proposal can also be adapted for different types of cancer through artificially functionalized lipid membranes with specific natural cell membranes, thus ensuring the release of drugs sensitive to the tumor microenvironment.

Doxorubicin is a broad-spectrum chemotherapeutic agent used alone or in combination to treat various solid tumors. However, it is a molecule that can cause cellular toxicity both acutely and chronically. This anthracycline can induce excessive formation of reactive oxygen species (ROS), senescence, morphological changes in healthy cells, and even lead to apoptosis [54]. The primary direct effect of doxorubicin is cardiotoxicity, and in response to this limitation, targeted liposomes have emerged, aiming to optimize the delivery of doxorubicin [55]. In response to this limitation, targeted liposomes have been synthesized, aiming to optimize the delivery of this therapeutic agent. This targeted approach aims to minimize unwanted side effects in healthy tissues while maximizing the effectiveness of cancer treatment. These liposomes act as smart delivery vehicles, selectively directing doxorubicin to cancer cells. In this way, they reduce systemic toxicity and improve the safety and efficacy of the treatment.

In the context of breast cancer, PEGylated liposomes functionalized with the incorporation of doxorubicin can contribute to the control of tumor growth. In this regard, ref. [56] functionalized liposomes with peptides targeting SREKA (Ser-Arg-Glu-Lys-Ala), whose identification is crucial for the effectiveness of cancer treatment, thus providing selective delivery of doxorubicin as a therapeutic agent. It was observed that inhibition in primary tumor growth and metastasis incidence was observed; moreover, it increased the survival rate of tumor-bearing mice.

PEGylation functionalization can also stabilize and target liposomes to the tumor site, increasing therapeutic efficacy and reducing systemic toxicity. Gold nanoparticles (AuNPs) are already used in therapies and diagnostics against different types of cancer [57]. Functionalized liposomes with PEGylated AuNPs confer a hybrid and targeted therapeutic strategy, in addition to pH and temperature-sensitive nucleolipids with functionalization. Sensitivity to pH and temperature changes in the tumor microenvironment allows for controlled and selective release of doxorubicin. At the same time, PEGylated AuNPs provide stability and targeting to liposomes, facilitating their accumulation in the tumor. This integrated approach aims to maximize the effectiveness of cancer treatment while minimizing the side effects associated with conventional chemotherapy, thus significantly contributing to the advancement of antitumor therapy [58].

Liposomes are also used in treating neurodegenerative diseases such as Alzheimer's and Parkinson's, not only for their properties but also for their ability to cross the blood-brain barrier (BBB) [59, 60] easily. Therefore, given the inherent complexity of neurological disorders such as Alzheimer's disease, which involve various mechanisms and affect multiple brain regions, the choice of liposomes offers a precise and targeted delivery approach [61].

Vitamin B12, although demonstrating anti-amyloidogenic properties *in vitro*, faces significant challenges, such as its high molecular weight and hydrophilicity, hindering its effective clinical application due to the difficulty of crossing the BBB. To overcome this limitation, liposomes were functionalized with transferrin (Tf), a protein that binds to transferrin receptors (TfRs) abundantly expressed in the endothelial cells of the BBB. This enabled receptor-mediated transcytosis, an effective strategy for crossing the BBB, increasing the efficiency and specificity of brain delivery against Alzheimer's disease. The developed nanosystem exhibited the capability to delay Aβ fibril formation and disaggregate mature fibrils, demonstrating its significant potential for Alzheimer's disease prevention and treatment [62].

Liposomes can selectively accumulate in inflamed sites, allowing them to deliver drugs specifically to inflamed tissues while sparing healthy ones. In this perspective, Ref. [63] investigated the potential use of dexamethasone (Dex), a corticosteroid with anti-inflammatory and immunosuppressive properties, in treating inflammatory lung diseases such as asthma, acute lung injury, and COVID-19. Despite its therapeutic benefits, prolonged systemic administration of Dex can result in adverse side effects. However, local pulmonary administration of corticosteroids, especially inhalation, is efficient and associated with better patient tolerance. The researchers developed three surface-modified liposomes containing Dex: Lip-PEG-Dex, Lip-PEGHA-Dex, and Lip-HA-Dex. These liposomes were designed to overcome the challenges of inhalation therapy, such as moderate deposition in the lower airways, toxicity to healthy lung tissues, and limited pulmonary retention time.

The inclusion of poly(ethylene glycol) (PEG) and/or hyaluronic acid (HA) in the nanoparticles aims to improve mucus penetration and target alveolar macrophages. *In vitro* assays conducted in RAW 264.7, macrophages indicated that liposomes containing HA showed more efficient targeting to activated macrophages. Furthermore, *in vivo* results in C57BL/6 J mice revealed more consistent efficacy of encapsulated Dex.

The nasal route has shown promise for insulin administration in treating diabetes. However, some barriers to drug absorption may compromise bioavailability. These barriers include thick nasal mucosa, rapid mucociliary clearance, and enzymatic degradation. Overcoming these barriers is essential to ensure the efficacy of nasal insulin administration and maximize its benefits in glycemic control for diabetic patients.

Liposomes loaded with insulin, when functionalized with cell-penetrating peptides (CPPs) such as TAT and Penetration (PNT), which act as promoters of drug penetration and absorption, promote lower release and permeation values through the nasal mucosa compared to liposomal systems without functionalization. This suggests that this behavior occurred due to electrostatic action. What happens is an interaction between insulin and CPPs and the consequent complexation between them, which reduces the incorporation of the drugs by the liposome and may inhibit the absorption-promoting activity of the peptide. However, liposomes loaded with insulin without functionalization with CPPs showed an increased permeability coefficient, suggesting that the developed system can optimize insulin absorption through the nasal route, which is an innovative aspect [64].

The advantages of liposome functionalization are extensive. Internally, the strategy allows for more effective and controlled drug loading, while external functionalization enables targeted and specific delivery, enhancing therapeutic efficacy. Both approaches contribute to reducing side effects and improving drug bioavailability. However, associated challenges include the complexity of the involved methodologies, careful and precise handling, and the need for rigorous control of experimental conditions to ensure consistent and reliable results. Additionally, issues related to the potential toxicity of the agents used need to be thoroughly assessed to ensure patient safety. These aspects underscore the importance of a multidisciplinary approach and careful preclinical evaluation for successfully developing these therapies. The functionalization of liposomes adds specific properties that go beyond their natural characteristics, allowing for targeted and effective drug delivery (**Table 2**). This reduces adverse effects such as side effects and toxicity and enhances therapeutic efficiency by increasing treatment precision. There are numerous possibilities for applying these particles, spanning areas such as oncology therapy, regenerative medicine, gene therapy, and the treatment of neurodegenerative diseases. Therefore, liposome functionalization, whether internal or external, represents a dynamic research field that offers opportunities for advances in personalized and precision drug delivery. Understanding these approaches' methodologies, advantages, and disadvantages is essential for shaping the future of drug therapy and diagnostics.

Type of functionalization	Type of bioactive	Applications	Study model	Author
Core-shell anti-phagocytosis-blocking repolarization-resistant membrane-fusogenic liposome (ARMFUL)	BLZ945 (colony stimulating factor receptor inhibitor), aCD7 (anti-CD47)	Adoptive cell therapy (ACT) effective against solid tumors	*in vitro:* M1 macrophages, B16F10 cells *in vivo:* C57BL/6 and BALB/c mice	[51]
Hybrid nanovesicle integrated into cancer cell membrane-derived liposome fusing with charge-reversed liposomal membrane	Phosphoglycerate mutase 1 and docetaxel (DXT) siRNA	Treatment of non-Small Cell Lung Cancer (NSCLC)	*in vitro*: A549 cells *in vivo*: BALB/c mice carrying A549 cell tumor	[52]
PEGylated liposomes targeting Ser-Arg-Glu-Lys-Ala (SREKA) peptide	Doxorubicin	Treatment of highly metastatic breast cancer	*in vitro*: 4 T1-Luc and NIH-3 T3 cells *in vivo*: BALB/c mice	[56]
Transferrin-functionalized liposomes	Vitamin B12 (VB12)	Treatment of Alzheimer's Disease	*in vitro*: cellulose dialysis membrane	[62]
Liposomes functionalized with cell-penetrating peptides (CPPs)	Insulin	Nasal administration	*in situ*: porcine nasal mucosa *in vitro*: Franz diffusion cell and synthetic cellulose acetate membrane	[64]
Liposomes containing nucleolipids sensitive to pH and temperature and functionalized with PEGylated AuNPs	Doxorubicin	Cancer therapy	*in vitro*: MDA-MB-231 and SK-OV-3 cells	[58]
Mito-liposome	Gold nanoparticles	Diagnostic method	*in vitro:* U2OS cells	[49]

Table 2.
Functionalization of liposomes.

DOI: http://dx.doi.org/10.5772/intechopen.1005237

5. Conclusion

In conclusion, the utilization of liposomes in drug delivery represents a pivotal advancement in pharmaceutical sciences, offering a versatile and efficient platform for various therapeutic applications. With their capability to encapsulate a wide range of drugs and provide controlled release, liposomes hold promise for optimizing therapy outcomes. Moreover, integrating nanomaterials with liposomes enhances formulation stability and facilitates precise drug targeting, further improving therapeutic efficacy. As research continues to explore the potential of liposomes, particularly in overcoming the challenges of drug bioavailability and stability, these nanostructures are poised to revolutionize drug delivery in the future. Integrating liposomes with nanomaterials stands at the forefront of innovation, offering tailored solutions to address complex therapeutic needs and ushering in a new era of precision medicine.

Acknowledgements

This work was supported by grants from CNPq, CAPES, FAPEAL, and RENORBIO.

Conflict of interest

The authors declare no conflict of interest.

Author details

Jerusa M. Oliveira[1,2], Rener M.F. Duarte[1,3], Samaysa de L. Lins[3,4], Lívia M.S. de Lima[1,2], Jéssica M. Pereira[1,2], Larissa I.M. de Almeida[1,2], Dhandara E. de L. Sampaio[1,2], Auana R. da S. Andrade[1,2], Isabella de O.F. de Sousa[5], Carlo J.F. Oliveira[5], Virmondes Rodrigues[5], Marcos V. da Silva[5], Foued Salmen Espindola[6], Fabiane C. de Abreu[4], Lucas Anhezini[2], Juliana Reis Machado e Silva[5] and Anielle Christine A. Silva[1,3]*

1 Strategic Materials Laboratory, Physics Institute, Federal University of Alagoas, Maceió, Alagoas, Brazil

2 Laboratory for in vivo Analysis of Toxicity and Neurodegenerative Diseases, Institute of Biological and Health Sciences, Federal University of Alagoas, Maceió, Alagoas, Brazil

3 Postgraduate Program of the Northeast Biotechnology Network (RENORBIO), Federal University of Alagoas, Maceió, Brazil

4 Laboratory of Electrochemistry and Analysis Microsystems, Institute of Chemistry and Biotechnology, Federal University of Alagoas, Maceió, Brazil

5 Department of Microbiology, Immunology and Parasitology, Institute of Biological and Natural Sciences, Federal University of Triângulo Mineiro, Uberaba, Minas Gerais, Brazil

6 Biochemistry and Molecular Biology Laboratory, Biotechnology Institute, Federal University of Uberlândia, Uberlândia, Minas Gerais, Brazil

*Address all correspondence to: aniellechristineas@gmail.com

References

[1] Kim EM, Jeong HJ. Liposomes: Biomedical applications. Chonnam Medical Journal. 2021;**57**(1):27-35. DOI: 10.4068/cmj.2021.57.1.27

[2] Sharma VK, Agrawal MK. A historical perspective of liposomes-a bio nanomaterial. Materials Today: Proceedings. 2021;**45**:2963-2966. DOI: 10.1016/j.matpr.2020.11.952

[3] Lombardo D, Kiselev MA. Methods of liposomes preparation: Formation and control factors of versatile nanocarriers for biomedical and nanomedicine application. Pharmaceutics. 2022;**14**:543. DOI: 10.3390/pharmaceutics14030543

[4] Guimarães D, Cavaco-Paulo A, Nogueira E. Design of liposomes as drug delivery system for therapeutic applications. International Journal of Pharmaceutics. 2021;**601**:120571. DOI: 10.1016/j.ijpharm.2021.120571

[5] Li T, Cipolla D, Rades T, Boyd BJ. Drug nanocrystallisation within liposomes. Journal of Controlled Release. 2018;**288**:96-110. DOI: 10.1016/j.jconrel.2018.09.001

[6] Nsairat H, Khater D, Sayed U, Odeh F, Bawab AA, Alshaer W. Liposomes: Structure, composition, types, and clinical applications. Heliyon, [S.L.]. maio, Elsevier BV; 2022;**8**(5):1-15. DOI: 10.1016/j.heliyon.2022.e09394

[7] Leo V, Maurelli AM, Giotta L, Catucci L. Liposomes containing nanoparticles: Preparation and applications. Colloids and Surfaces B: Biointerfaces. [S.L.], Vol. 218. out. 2022. p. 112737. DOI: 10.1016/j.colsurfb.2022.112737

[8] Niesman MR, Khoobehi B, Magin RL, Webb AG. Liposomes and diagnostic imaging: The potential to visualize both structure and function. Journal of Liposome Research. 1994;**4**(2):741-768. DOI: 10.3109/08982109409018597

[9] Musielak M, Potoczny J, Boś-Liedke A, Kozak M. The combination of liposomes and metallic nanoparticles as multifunctional nanostructures in the therapy and medical imaging—A review. International Journal of Molecular Science. 2021;**22**:6229. DOI: 10.3390/ijms22126229

[10] Kim EM, Jeong HJ. Liposomes: Biomedical applications. Chonnam Medicine Journal. 2021;**57**(1):27-35. DOI: 10.4068/cmj.2021.57.1.27

[11] Chen LJ, Yang CX, Yan XP. Liposome-coated persistent luminescence nanoparticles as luminescence trackable drug carrier for chemotherapy. Analytical Chemistry. 2017;**89**(13):6936-6939. DOI: 10.1021/acs.analchem.7b01397

[12] Sun SK, Wang HF, Yan XP. Engineering persistent luminescence nanoparticles for biological applications: From biosensing/bioimaging to theranostics. Accounts of Chemical Research. 2018;**51**(5):1131-1143. DOI: 10.1021/acs.accounts.7b00619

[13] Chen CS, Yao J, Durst RA. Liposome encapsulation of fluorescent nanoparticles: Quantum dots and silica nanoparticles. Journal of Nanoparticle Research. 2006;**8**(6):1033-1038. DOI: 10.1007/s11051-006-9142-1

[14] Yang J, He S, Hu Z, Zhang Z, Cao C, Cheng Z, et al. In vivo multifunctional fluorescence imaging using liposome-coated lanthanide nanoparticles in near-infrared-II/IIa/IIb windows. Nano

Today. 2021;**38**:101120. DOI: 10.1016/j.nantod.2021.101120

[15] Lacerda S, Delalande A, Eliseeva SV, Pallier A, Bonnet CS, Szeremeta F, et al. Doxorubicin-sensitized luminescence of NIR-emitting ytterbium liposomes: Towards direct monitoring of drug release. Angewandte Chemie. 2021;**133**(44):23766-23769. DOI: 10.1002/ange.202109408

[16] Musielak M, Potoczny J, Boś-Liedke A, Kozak M. The combination of liposomes and metallic nanoparticles as multifunctional nanostructures in the therapy and medical imaging-a review. International Journal of Molecular Sciences. 2021;**22**(12):6229. DOI: 10.3390/ijms22126229

[17] Skóra B, Szychowski KA, Gmiński J. A concise review of metallic nanoparticles encapsulation methods and their potential use in anticancer therapy and medicine. European Journal of Pharmaceutics and Biopharmaceutics. 2020;**154**:153-165. DOI: 10.1016/j.ejpb.2020.07.002

[18] Shrestha B, Tang L, Romero G. Nanoparticles-mediated combination therapies for cancer treatment. Advanced Therapeutics. [S.L.]. set. Wiley; 2019;**2**(11,16):1-7. DOI: 10.1002/adtp.201900076

[19] Makowski M, Silva ÍC, Pais do Amaral C, Gonçalves S, Santos NC. Advances in lipid and metal nanoparticles for antimicrobial peptide delivery. Pharmaceutics. 2019;**11**.(11):588. DOI: 10.3390/pharmaceutics11110588

[20] Musielak M, Potoczny J, Boś-Liedke A, Kozak M. The combination of liposomes and metallic nanoparticles as multifunctional nanostructures in the therapy and medical imaging—A review. International Journal of Molecular Science. 2021;**22**:6229. DOI: 10.3390/ijms22126229

[21] Stoltzfus RJ. Iron deficiency: Global prevalence and consequences. Food and Nutrition Bulletin. 2003;**24**(4 Suppl):S99-S103. DOI: 10.1177/15648265030244S206

[22] Bjørklund G, Semenova Y, Hangan T, Pen JJ, Aaseth J, Peana M. Perspectives on iron deficiency as a cause of human disease in global public health. Current Medicinal Chemistry. 2024;**31**(12):1428-1440. DOI: 10.2174/0929867330666230324154606

[23] Fathy MM, Fahmy HM, Balah AMM, Mohamed FF, Elshemey WM. Magnetic nanoparticles-loaded liposomes as a novel treatment agent for iron deficiency anemia: In vivo study. Life Sciences. 2019;**234**:116787. DOI: 10.1016/j.lfs.2019.116787

[24] Dorjsuren B, Chaurasiya B, Ye Z, Liu Y, Li W, Wang C, et al. Cetuximab-coated thermo-sensitive liposomes loaded with magnetic nanoparticles and doxorubicin for targeted EGFR-expressing breast cancer combined therapy. International Journal of Nanomedicine. 2020;**15**:8201-8215. DOI: 10.2147/ijn.s261671

[25] Anilkumar TS, Lu YJ, Chen JP. Optimization of the preparation of magnetic liposomes for the combined use of magnetic hyperthermia and photothermia in dual magneto-photothermal cancer therapy. International Journal of Molecular Science. 2020;**21**:5187. DOI: 10.3390/ijms21155187

[26] Kopeckova K, Eckschlager T, Sirc J, Hobzova R, Plch J, Hrabeta J,

et al. Nanodrugs used in cancer therapy. Biomedical papers of the Medical Faculty of the University Palacky, Olomouc, Czech Republic. 2019;**163**(2):122-131. DOI: 10.5507/bp.2019.010

[27] Wang Z, Li J, Lin G, He Z, Wang Y. Metal complex-based liposomes: Applications and prospects in cancer diagnostics and therapeutics. Journal of Controlled Release. 2022;**348**:1066-1088. DOI: 10.1016/j.jconrel.2022.06.012

[28] Hussein HA, Abdullah MA. Novel drug delivery systems based on silver nanoparticles, hyaluronic acid, lipid nanoparticles and liposomes for cancer treatment. Applied Nanoscience. 2022;**12**:3071-3096. DOI: 10.1007/s13204-021-02018-9

[29] Desai N, Momin M, Khan T, Gharat S, Ningthoujam RS, Omri A. Metallic nanoparticles as drug delivery system for the treatment of cancer. Expert Opinion on Drug Delivery. 2021;**18**(9):1261-1290. DOI: 10.1080/17425247.2021.1912008

[30] Ahmed SA, Gaber MH, Salama AA, Ali SA. Efficacy of copper nanoparticles encapsulated in soya lecithin liposomes in treating breast cancer cells (MCF-7) in vitro. Scientific Reports. 2023;**13**:15576. DOI: 10.1038/s41598-023-42514-2

[31] Charest G, Tippayamontri T, Shi M, Wehbe M, Anantha M, Bally M, et al. Concomitant Chemoradiation therapy with gold nanoparticles and platinum drugs co-encapsulated in liposomes. International Journal of Molecular Science. 2020;**21**:4848. DOI: 10.3390/ijms21144848

[32] Karchemski F, Zucker D, Barenholz Y, Regev O. Carbon nanotubes-liposomes conjugate as a platform for drug delivery into cells. Journal of Controlled Release. 2012;**160**(2):339-345. DOI: 10.1016/j.jconrel.2011.12.037

[33] Contreras M, Torres C, Villarroel I, Rozas R. Molecular dynamics assessment of doxorubicin–carbon nanotubes molecular interactions for the design of drug delivery systems. Structural Chemistry. 2019;**30**:369-384. DOI: 10.1007/s11224-018-1210-5

[34] Silva ACA, Alvin EA, Jesus LS, França CCL, Silva MPG, Lins SL, et al. NanoBioSensors: From electrochemical sensors improvement to theranostic applications. In: Asadpour V, Karakus S, editors. Biosignal Processing. Biomedical Engineering. [S.L.]. IntechOpen; 21 dez. 2022. DOI: 10.5772/intechopen.102552

[35] Liu J, Ma C, Shi S, Liu H, Wen W, Zhang X, et al. A general controllable release amplification strategy of liposomes for single-particle collision electrochemical biosensing. Biosensors and Bioelectronics. 2022;**207**:114182. DOI: 10.1016/j.bios.2022.114182

[36] Ahmed HM, Nabavi S, Behzad S. Herbal drugs and natural products in the light of nanotechnology and nanomedicine for developing drug formulations. Mini-Reviews in Medicinal Chemistry. 2021;**21**(3):302-313. DOI: 10.2174/1389557520666200916143240

[37] Shishir MRI, Karim N, Gowd V, Zheng X, Chen W. Liposomal delivery of natural product: A promising approach in health research. Trends in Food Science & Technology. 2019;**85**:177-200. DOI: 10.1016/j.tifs.2019.01.013

[38] Sogut O, Sezer UA, Sezer S. Liposomal delivery systems for herbal extracts. Journal of Drug Delivery

Science and Technology. 2021;**61**:102147. DOI: 10.1016/j.jddst.2020.102147

[39] Miere F, Vicas SI, Timar AV, Ganea M, Zdrinca M, Cavalu S, et al. Preparation and characterization of two different liposomal formulations with bioactive natural extract for multiple applications. Processes. 2021;**9**:432. DOI: 10.3390/pr9030432

[40] Natsaridis E, Gkartziou F, Mourtas S, Stuart MCA, Kolonitsiou F, Klepetsanis P, et al. Moxifloxacin liposomes: Effect of liposome preparation method on physicochemical properties and antimicrobial activity against staphylococcus epidermidis. Pharmaceutics. 2022;**14**:370. DOI: 10.3390/pharmaceutics14020370

[41] Saddiqi ME, Kadir AA, Abdullah FFJ, Zakaria MZAB, Banke IS. Preparation, characterization and in vitro cytotoxicity evaluation of free and liposome-encapsulated tylosin. OpenNano. 2022;**8**:100108. DOI: 10.1016/j.onano.2022.100108

[42] Rakshit M, Darwitan A, Muktabar A, Das P, Nguyen LTH, Cao Y, et al. Anti-inflammatory potential of simvastatin loaded nanoliposomes in 2D and 3D foam cell models. Nanomedicine: Nanotechnology, Biology and Medicine. 2021;**37**:102434. DOI: 10.1016/j.nano.2021.102434

[43] Peer D, Karp J, Hong S, et al. Nanocarriers as an emerging platform for cancer therapy. Nature Nanotechnology. 2007;**2**:751-760. DOI: 10.1038/nnano.2007.387

[44] Wu J. The enhanced permeability and retention (EPR) effect: The significance of the concept and methods to enhance its application. Journal of Personalized Medicine. 2021;**11**:771. DOI: 10.3390/jpm11080771

[45] Murugesan K, Srinivasan P, Mahadeva R, Gupta CM, Haq W. Tuftsin-bearing liposomes co-encapsulated with doxorubicin and curcumin efficiently inhibit EAC tumor growth in mice. International Journal of Nanomedicine. 2020;**15**:10547-10559. DOI: 10.2147/IJN.S276336

[46] Saka R, Chella N. Nanotechnology for delivery of natural therapeutic substances: A review. Environmental Chemistry Letters. 2021;**19**:1097-1106. DOI: 10.1007/s10311-020-01103-9

[47] Ramli NA, Ali N, Hamzah S, Yatim NI. Physicochemical characteristics of liposome encapsulation of stingless bees' propolis. Heliyon. 2021;7(4):e06649. DOI: 10.1016/j.heliyon.2021.e06649

[48] Melchior S, Codrich M, Gorassini A, Mehn D, Ponti J, Verardo G, et al. Design and advanced characterization of quercetin-loaded nano-liposomes prepared by high-pressure homogenization. Food Chemistry. 2023;**428**:136680. DOI: 10.1016/j.foodchem.2023.136680

[49] Sardarabadi H, Chafai DE, Gheybi F, Sasanpour P, Rafii-Tabar H, Cifra M. Enhancement of the biological autoluminescence by mito-liposomal gold nanoparticle nanocarriers. Journal of Photochemistry and Photobiology B: Biology. 2020;**204**:111812. DOI: 10.1016/j.jphotobiol.2020.111812

[50] Tenchov R, Bird R, Curtze AE, Zhou Q. Lipid nanoparticles—from liposomes to mRNA vaccine delivery, a landscape of research diversity and advancement. ACS Nano. 2021;**15**(11):16982-17015. DOI: 10.1021/acsnano.1c04996

[51] Zheng C, Zhong Q, Yi K, Kong H, Cao F, Zhuo C, et al. Anti–phagocytosis-blocking repolarization-resistant membrane-fusogenic liposome

(ARMFUL) for adoptive cell immunotherapy. Science Advances. 2023;**9**(32):1-3. DOI: 10.1126/sciadv.adh2413

[52] Zhang W, Gong C, Chen Z, Li M, Li Y, Gao J. Tumor microenvironment-activated cancer cell membrane-liposome hybrid nanoparticle-mediated synergistic metabolic therapy and chemotherapy for non-small cell lung cancer. Journal of Nanobiotechnology. 2021;**19**:339. DOI: 10.1186/s12951-021-01085-y

[53] Gridelli C, Rossi A, Carbone DP, Guarize J, Karachaliou N, Mok T, et al. Non-small-cell lung cancer. Nature Reviews Disease Primers. 2015;**21**(1):15009. DOI: 10.1038/nrdp.2015.9

[54] Cappetta D, Rossi F, Piegari E, Quaini F, Berrino L, Urbanek K, et al. Doxorubicin targets multiple players: A new view of an old problem. Pharmacological Research. 2018;**127**:4-14. DOI: 10.1016/j.phrs.2017.03.016

[55] Moosavian SA, Sahebkar A. Aptamer-functionalized liposomes for targeted cancer therapy. Cancer Letters. 2019;**448**:144-154. DOI: 10.1016/j.canlet.2019.01.045

[56] Vári B, Dókus L, Borbély A, Gaál A, Vári-Mező D, Ranđelović I, et al. SREKA-targeted liposomes for highly metastatic breast cancer therapy. Drug Delivery. 2023;**30**:1. DOI: 10.1080/10717544.2023.2174210

[57] D'acunto M, Cioni P, Gabellieri E, Presciuttini G. Exploiting gold nanoparticles for diagnosis and cancer treatments. Nanotechnology. 2021;**32**(19):192001. DOI: 10.1088/1361-6528/abe1ed

[58] García MC, Calderón-Montaño JM, Rueda M, Longhi M, Rabasco AM, López-Lázaro M, et al. PH-temperature dual-sensitive nucleolipid-containing stealth liposomes anchored with PEGylated AuNPs for triggering delivery of doxorubicin. International Journal of Pharmaceutics. 2022;**619**:121691. DOI: 10.1016/j.ijpharm.2022.121691

[59] Spuch C, Navarro C. Liposomes for targeted delivery of active agents against neurodegenerative diseases (Alzheimer's Disease and Parkinson's Disease). Journal of Drug Delivery, [S.L.], V. 2011. Hindawi Limited; 13 dez. 2011. p. 1-12. DOI: 10.1155/2011/469679

[60] Khot KB, Gopan G, Bandiwadekar A, Jose J. Current advancements related to phytobioactive compounds based liposomal delivery for neurodegenerative diseases. Ageing Research Reviews. 2023;**83**:101806. DOI: 10.1016/j.arr.2022.101806

[61] Pandian SRK, Vijayakumar KK, Murugesan S, Kunjiappan S. Liposomes: An emerging carrier for targeting Alzheimer's and Parkinson's diseases. Heliyon. 2022;**8**(6):e09575. DOI: 10.1016/j.heliyon.2022.e09575

[62] Andrade S, Ramalho MJ, Loureiro JÁ, Pereira MC. Transferrin-functionalized liposomes loaded with vitamin VB12 for Alzheimer's disease therapy. International Journal of Pharmaceutics. 2022;**626**:122167. DOI: 10.1016/j.ijpharm.2022.122167

[63] Japiassu KB, Fay F, Marengo A, Mendanha AS, Cailleau C, Louaguenouni Y, et al. Hyaluronic acid-conjugated liposomes loaded with dexamethasone: A promising approach for the treatment of inflammatory diseases. International Journal of Pharmaceutics. 2023;**639**:122946. DOI: 10.1016/j.ijpharm.2023.122946

[64] Von Zuben ES, Eloy JO, Araujo VHS, Gremião MPD, Chorilli M. Insulin-loaded liposomes functionalized with cell-penetrating peptides: Influence on drug release and permeation through porcine nasal mucosa. Colloids and Surfaces A: Physicochemical and Engineering Aspects. 2021;**622**:126624. DOI: 10.1016/j.colsurfa.2021.126624

Chapter 5

Structure-Function Relationships in the Modification of Liposomes for Targeted Drug Delivery in Infectious Diseases

Palesa Pamela Seele

Abstract

The introduction of liposomes has caused a paradigm shift in medicine, offering novel solutions to problems that are ancient to the drug discovery and development for HIV, TB, and malaria. These are the three deadliest infectious diseases that are endowed with complex pathophysiological and biological mechanisms that allow them to thrive in their hosts through escaping the immune system and capturing key pathways. Disease heterogeneity and lack of suitable models to replicate the disease states make compounds the poor pharmacokinetic issues associated with these diseases. Liposomes are lipid-based nanocarriers that are employed for drug formulations, preservation, and storage. Importantly, they can be tailored for targeted and controlled release. Structure–function relationships are crucial to consider in liposome design as they affect key interactions between the carrier drug and the target cell, which impact on drug release, cellular uptake, bioavailability, biodistribution, and toxicity. Herein, lipid composition, size, lamellarity, zeta potential/charge as well as surface modification with cholesterol, PEG, peptides, and antibodies are discussed with respect to selectivity in targeting diseased cells. The role of computational tools in expediting the liposome technology is reviewed, highlighting the impact of forces of interaction between biomolecules and the conditions of the environment.

Keywords: liposomes, drug delivery systems (DDS), human immunodeficiency virus (HIV), tuberculosis (TB), malaria

1. Introduction

The phenomenon of "pandemic readiness" cannot be overemphasized, and through innovative management and preventive tools, the global manifestation of infectious diseases can be overcome. It is of paramount importance that these measures, which encompass therapeutics, vaccines, and diagnostics be developed with consideration for sustainability, biocompatibility, and biodegradability—safety for human and animal consumption, and eco-friendliness are imperative. Nanotechnology has provided an invigorating landscape for exploring solutions to ancient problems that still persist in modern-day medicine, hence the coining of the term nanomedicine. The introduction of liposomes has had a paradigm shift in medicine; their derivation from biological molecules has implored their service as valuable nano-vehicles in various bio-applications. Not only were they the first nano-medicine to be applied to human patients but also their success in clinical applications is undoubtedly apparent [1, 2].

Infectious diseases (also communicable diseases) are caused by pathogenic microorganisms, which are transmitted between animals through direct and indirect contact [3]. Pathogenic microorganisms have existed for centuries evolving mechanisms that would facilitate their evasion of the human immune system, including human-borne interventions that were developed to curb their spread such as vaccines, therapeutic drugs, and diagnostic tools. Low-resourced countries are the most vulnerable and burdened with infectious diseases which accounted for the highest mortality rate of about 2.5 million in 2020 from malaria, human immunodeficiency virus (HIV), and tuberculosis (TB) [3–5]. Malaria, TB, and HIV/AIDS were the most prominent infectious diseases in WHO's top 10 leading causes of death in low-income countries for the year 2019 and are the three deadliest infectious diseases globally [5, 6]. This has necessitated the development of cost-effective methods that are efficient and accessible. The heterogeneity in the pathophysiology of these diseases, their canny mechanisms of capturing, and exploiting their hosts' systems allow them to thrive while making it difficult to formulate drugs that are nontoxic for the host. Moreover, models that can replicate the disease states do not exist, which further complicates drug validation processes. Poor pharmacokinetics and drug toxicity are thus prominent issues that require novel approaches.

Liposomes are broadly used in industry, including in the pharmaceutical, nutraceutical as well as other applications in biotechnology [7]. They act as cell membrane models and as carriers for targeted delivery systems, such as drugs, vaccines, and various biomolecules: nucleic acids, proteins, peptides, lipids, carbohydrates, antibiotics, dyes, antioxidants, and enzymes [7].

1.1 The history and biochemistry of liposomes

Liposome is a portmanteau of Greek words "lipos" and "soma" meaning "fat" and "body," respectively, which in essence describes their amphiphilic phospholipid building blocks [8]. Consequently, they are vesicle-forming, with the phospholipid bilayer enclosing an aqueous core, which dissolves hydrophilic compounds, while the lipid layer entraps lipophilic compounds [7]. When mixed with therapeutic compounds, lipids can transform into NP that entrap and protect their cargo, releasing them upon encounter with their target cells. The structure is depicted in **Figure 1**.

Liposomes were first evidenced using electron microscopy in the 1960s by Alec Bangham while studying phospholipids and blood clotting factors [8]. The history of liposomes is outlined in **Figure 2**. Major strides were reported between the late 1980s and 1990s when the anticancer drug doxorubicin was successfully encapsulated and later approved in 1995 - Doxil® [10]. These were important discoveries where liposomes demonstrated their ability to reduce doxorubicin-induced cardiotoxicity [11, 12] and increase blood residence [13]. The upscaling of liposome production, introduction of new lipid raw material, and the tailorability of the lipid material were some of the major progressive events in the history of liposomes as drug delivery

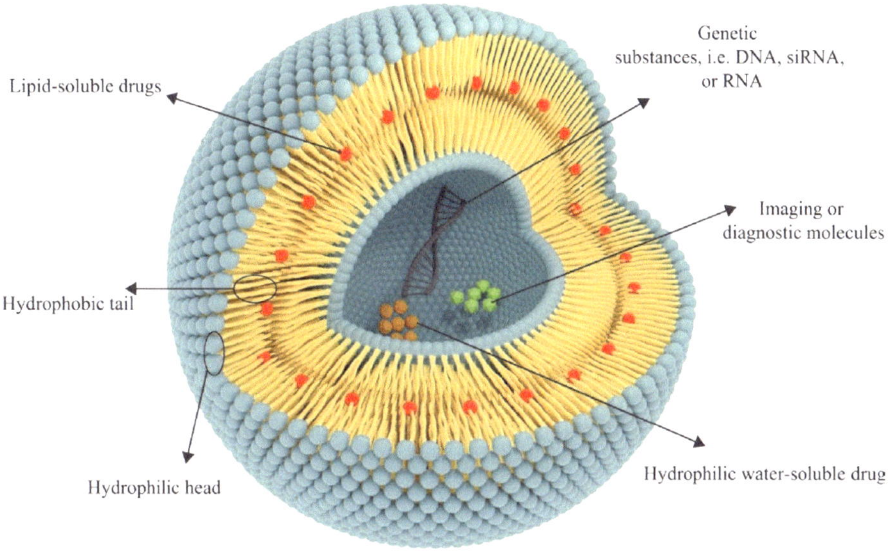

Figure 1.
Structure of a liposome. (The figure has been reproduced from a review by Rommasi and colleagues [9], with permission from SpringerOpen and open access from the creative commons CC-BY license).

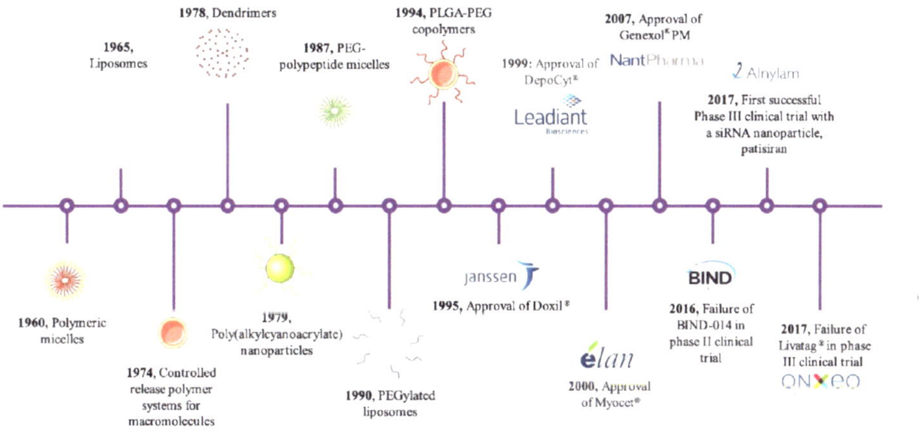

Figure 2.
Timeline history of liposomes. (The figure was reproduced from Elsevier [10], with open access from the creative commons CC-BY license).

systems [8]. To date, a variety of lipid formulations are used from natural and/or synthetic lipids and surfactants [14].

The size, lamellarity, surface charge, zeta potential, lipid composition, and organization of liposomes are imperative properties in their efficacy, playing key roles in their interaction with the cell, blood residence or half-life, tissue permeability, and final fate *in vivo* [14]. The morphology and physicochemical properties of liposomes, including the dimension and lamellarity dictates their classification and preferred applications [7, 14]. This will be discussed in detail in the sections to follow.

Cholesterol is mandatory to the liposome formulation, making up 30-50% of the membrane on average, which is higher than any membrane component [15–17]. It intercalates readily in the membrane core due to its amphipathic properties resembling the phospholipids, with a predominantly hydrophobic structure. Cholesterol has a polar hydroxyl (OH) group that connects to four fused hydrocarbon rings that are, in turn, linked to a branched chain of eight carbon residues [18]. In essence, the arrangement of cholesterol in the liposomes is not coincidental, and it is positioned perpendicularly to the plane of the bilayer, wherein the polar OH group, that is, the 3β-OH is in proximity with the phosphate and carbonyl of the glycerol ester linkage of the phospholipids, allowing the potential formation of hydrogen bonds (H-bonds), which are considered the most important of interactions [19]. Hydrogen bonds can also form through water bridges between the cholesterol OH group and any acceptor or donor group from the phospholipid, and finally, the oxygen of the OH group and phospholipids CH_3 group [19, 20]. However, the rigidity or increased stability of the phospholipid bilayer is not only endowed by the hydrogen bonding but also the aromatic rings of cholesterol, which contain major and minor grooves, and the rings are proposed to reduce the flexibility of the aliphatic carbon chain of the phospholipids [21]. The reduced membrane permeability induced by cholesterol also owes to the hydrophobic effect between the lipid hydrocarbon chains and cholesterol, which is imperative for conferring rigidity [22]. Through these interactions, cholesterol is capable of inducing morphological changes that modify the bilayer stability, curvature, membrane fluidity, diffusion rates of proteins and lipids, and permeability, which directly impact drug leakage.

Phospholipid derivatives: Fluidity and charge of the membrane are greatly dependent on the phospholipid derivatives. These can be in the form of phosphatidylglycerol, phosphatidylserine, phosphatidylcholine, or phosphatidylethanolamines [23]. Permeability or drug leakage can be controlled by the type of lipid used; whereas saturated acyl chain lipids confer rigidity and impermeability, unsaturated lipids tend to be more permeable [23]. The types of lipid composition and effect on drug delivery systems are discussed in detail in the below sections, with specific references to the disease.

Size and lamellarity: Liposomes assume a spherical structure, which can range in size from nanometers to micrometers; however, in medicine, nanoliposomes have been more acceptable [14], providing high-surface areas that afford the loading of high concentrations of drugs and vaccines. Size is an important effector in drug loading and release, bioavailability, biodistribution, mucoadhesion, and cellular uptake [24]. Particle size as well as the polydispersity index (PDI) are essential for cellular uptake—a process that mainly occurs *via* an endocytosis-dependent manner. Efficiency in the systemic delivery of drugs and uptake by tissues is dependent on the capillary perfusion of tissues, wherein the efficiency in cargo exchange depends on the size of the cargo delivered and of the membranous fenestration [24].

The aerosol delivery of drugs is favored for diseases of the lung, wherein the harsh systemic conditions are avoided to enable proximal localization of the drug to its target. The aerodynamic volume of the liposome-encapsulated system becomes important in the pulmonary localization of the drug, wherein drugs with a mass median aerodynamic diameter (MMAD) of 3 μm have an estimated 50-60% chance of localizing in the alveoli and 80% chance of residence in the lower airways [25]. In comparison, the oropharyngeal and large conducting airways regions are mostly adsorbed with larger particles with a size range of 5–10 μm [25]. The impermeability of the blood-brain barrier (BBB) has imposed challenges in the development of efficacious drugs targeting the brain tissue requiring innovative methods of enhancing their bioavailability and biodistribution. Liposomes ≤100 nm have been reported to improve the delivery of drugs to the brain and central nervous system for cancer therapy [24]. Insight from such studies can facilitate the design of liposome-based delivery systems for the treatment of TB meningitis.

Malarial drugs targeting the hepatic stages of the disease also have to consider liposome size as it determines the hepatic uptake and accumulation where capillary exchange of particles ≤150 nm is acceptable [24]. In addition to size, the curvature of particles impacts on the internalization in the Kupffer cells. While particles greater than 1 μm with elongated shapes enhance particle to membrane contacts, the high-aspect rations often inhibit membrane spread over the smaller dimensions, limiting the completion of phagocytosis. In contrast, with particles that are less than 1 μm, the shape determines the rate and pathway of internalization [26]. Delivery to the parenchymal cells is achieved by particles that are less than 50 nm in size.

To reiterate that size matters, liposomes are classified according to their size-lamellarity, single unilamellar vesicles (SUV), medium unilamellar vesicles (MUV), large unilamellar vesicles (LUV), and giant unilamellar vesicles (GUV). Further, annotations are given in terms of lamellarity, which denotes the number of phospholipid bilayers, either being unilamellar, oligolamellar, or multilamellar [14]. This makes liposomes differ in their functionality. The LUV liposomes are ideal for encapsulating hydrophilic cargo in higher amounts due to their larger entrapped aqueous volume, making them indifferent to the phospholipid types making them up [27].

Zeta potential and charge: Zeta potential is the potential difference between the layer of fluid that is stationary, remaining attached to the surface of a dispersed medium, and the dispersion medium; it can be used as an indirect measure of surface charge [27]. The any extreme shifts can result in structural changes, stability, and functionality of the liposomes, and temperature and viscosity are also effectors to consider [27, 28]. Surface functionalization of liposomes with (bio)molecules such as antibodies, nucleic acids, proteins, and dyes depends on their zeta potential as well as overall charge and polarity. Liposomes are considered stable at a minimum zeta potential of ±30 mV, with extremely positive or negative values indicating that colloidal molecules repel each other displaying dispersed colloidal particle [28]. A nano-quantitative structure-property relationships (QSPR) model study reports that the hydrophilic-lipophilic balance (HLB) and enthalpy of formation are structural features affecting the zeta potential; wherein a lower HLB value displayed higher zeta values, which means, the more lipophilic the higher the stability, and conversely, high enthalpy of formations was associated with lower zeta potential values and instability [29].

With respect to charge, electrostatic interactions between the cargo and vehicle are key players in the efficiency of encapsulation and delivery [28]. Cationic liposomes are predominantly used over their anionic counterparts due to their ease of binding to bacterial membranes and increased drug delivery. However, issues of toxicity have been evident and successfully circumvented — this is discussed in detail in the latter sections. Peetla and co-workers describe four types of interactions between the vehicle and cargo: hydrophilic and electrostatic interactions with a bilayer, water-soluble and non-bilayer interacting, non-water-soluble and non-bilayer interactions, and lastly, lipophilic and bilayer interacting [30].

The partitioning of drugs between the aqueous core and the lipid bilayer is a common occurrence for amphiphilic drugs, which is influenced by changes in the pH of the aqueous core. The non-bilayer is the aqueous core of the liposome, where the water-soluble drugs are encapsulated. Thus, a total entrapped aqueous volume that is large allows for higher concentrations of drug loading. In contrast, non-water-soluble drugs that are non-bilayer interacting have very poor encapsulation efficiencies, unless molecules with functionally compatible moieties are attached to the drug and/or the liposome is reformulated with a different type of lipid and/or accessory molecules. The length of the acyl chain of the phospholipid together with fluidity impacts the loading capacity or intercalation of lipophilic drugs on the bilayer. Incorporation of cholesterol can then be used to manipulate the fluidity of the lipid membrane [27, 30].

2. Liposomes as drug delivery systems in HIV, TB, and malaria

Drug delivery systems (DDS) are pivotal in managing and treating diseases, acting as vehicles for drug formulations, preservation, storage, and targeted release—usually entailing the encapsulation of the cargo. Administration of local and systemic drugs is usually through oral methods and various other routes, including cutaneous, nasal, ophthalmic, anal, vaginal as well as buccal and sublingual routes [31–33]. Once introduced into the system, the exposure of drugs to abrasive physicochemical environments and antagonizing physiological conditions pose serious limitations to their pharmacokinetics. Conditions that promote the decomposition or premature degradation of drugs and their excretion are associated with extreme pH and temperature, hydrolyzing enzymes, free radicals, and others. Thus, novel designs in DDS have to counteract poor selectivity, uncontrollable drug release, low solubility, short plasma residence, and premature drug metabolism and excretion, which are effectors of low drug efficacy, overdosage, high levels of toxicity, and patient incompliance. In principle, high efficacy, automation, precision as well as high biocompatibility and biodegradability are defining properties of a successful DDS [33]. These qualities are mitigating factors to the trend in acquired drug resistance of microbes, which is a result of overcompensation of poor pharmacokinetics by misuse and overuse of antimicrobials [34].

Liposomes are lipid-based NP, a category of DDS where the most FDA-approved nanomedicine originates [35] — warranted by their versatility in loading capacity, which enables the encapsulation of amphipathic, hydrophilic, and lipophilic bioactive compounds in their aqueous core and intercalation within the lipid leaflets, respectively. Their favorability also owes to their ability to self-assemble and carry high loads, amenability to functionalization, high biocompatibility, and bioavailability [35]. Structure-function relations are inherent in endowing these properties, which need to be unpacked and comprehended. The liposomes' manner of interaction with cells differs depending on the respective physicochemical properties; these interactive pathways occur via lipid exchange, liposome fusion, adsorption, and lastly and most vital is by endocytosis [36].

2.1 Human immunodeficiency virus (HIV)

Since the approval of the first antiviral drug in 1963, drug discovery against viruses has evolved simultaneously with genomic sequencing and the structure-function-based drug design. This era also brought up interest and effort in deconvolution of the HIV lifecycle, which shed insight into how the virus exploits the hosts' resources for propagation—an attribute to the difficulty in developing viral therapy without eliciting toxicity in bacteria [37, 38]. This led to the discovery of other drugs strategically targeting the different stages of the life cycle surpassing antibacterial drugs with better specificity, and with respect to HIV, the discovery of the protease was one of major breakthroughs. However, some improvement is warranted—existing side effects, and continued innovation is important to safeguard against drug resistance as well as achieve a cure. Moreover, comorbidities and polypharmacy are compounding factors of drug-drug interactions that lead to poor pharmacokinetics and pharmacodynamics, especially in adults [39]. To date, the highly active antiretroviral therapy (HAART) is still in effect, although efficient, some challenges are ongoing, including low oral bioavailability due to premature metabolism and degradation in the gut and virus' ability to evade therapy since it resides in unattainable anatomical and cellular reservoirs – the minimum drug concentration is not affected; toxicity is caused by high dosages as compensation to the low half-life, leading to noncompliance in patients [40]. Africa is still the most burdened, accounting for more than 65% of people living with HIV, but with a remarkable 38 and 51% decline in new incidences and mortality compared to the year 2010, respectively.

Azidothymidine (AZT), also known as zidovudine (ZDV), is a reverse transcriptase inhibitor and the first anti-HIV drug to be approved in 1987, and it is also the first to be encapsulated by liposomes from work that was initiated in 1991 by Phillips and team [41]. Conceptually, the work was built on the knowledge that the phagocytic system of the reticuloendothelial system (REM), and the macrophages, in particular, are highly active in the uptake of liposomes [42, 43]. The liposome-encapsulated AZT was able to reduce hematopoietic toxicity, enhancing antiretroviral activity and prophylaxis in mice infected with HIV [41], and the distearoylphosphatidylcholine/ dimyristoylphosphatidylglycerol (DSPC/ DMPG) liposomes showed three-fold higher retention time versus dimyristoylphosphatidylglycerol/dimyristoylphosphatidylglycerol (DPPC/DMPG) liposomes. Given that there are limitations in liposome-encapsulation of ARV, such as poor hydrophilic loading capacity, instability associated with the physical and biological make-up, poor scale-up and high cost, short shelf-life, and toxicity — through modern technology and by varying lipid properties such as compositions, size, and charge, these challenges may be overcome. AZT is amphiphilic, thus, leakage is caused by the split affinity between the liposome aqueous core and the hydrophobic lipid bilayer [42]. Jin and co-workers enhanced liposome retention of AZT by encapsulating the lipophilic prodrug form which intercalates better with the lipid bilayer [43]. Subsequently, the encapsulated prodrug displayed a half-life that was enhanced in the rat model. Modifying liposomes with polyethylene glycol (PEG), so-called PEGylation, is a well-accepted strategy that not only increases hydrophilicity but acts also as a steric barrier—preventing opsonins in the serum from tagging the liposome cargo for clearance [3]. Consequently, PEGylation increases retention time, improving bioavailability of the drug; this has been excellently demonstrated in the PEGylated-liposome encapsulating doxorubicin, commercially Doxil® — the first liposome-encapsulated drug to be approved and used commercially in treating AIDS-related Kaposi's sarcoma [44]. Saquinavir, a HIV protease inhibitor, showed more stability in protein supplemented media and a more sustainable release when encapsulated in PEGylated liposomes compared to free drug and non-PEG liposomes. The PEGylated liposomes were also less toxic to Jurkat T-cells [40].

The localization of encapsulated drugs on the liposome is crucial to consider as it affects drug leakage and can be exploited for controlled release. Saquinavir and nevirapine were observed to concentrate at different regions of the nanocarrier when they were simultaneously loaded; moreover, their release was in a timely manner, wherein the saquinavir was dominant during the later phase and nevirapine at the early stages [45]. Simultaneous release of the drugs is an approach that was aimed at circumventing the metabolism of saquinavir prior reach of the effective concentration. The acyl group of the bilayer allows association of the more hydrophobic nevirapine, while saquinavir localizes to the aqueous core causing its delay in release. The composition of liposome and surface modification affects entrapment of the drugs—PEGylation increases the hydrophilicity of the nanocarrier, significantly decreasing the encapsulation of nevirapine by 45%, while a slight increase was observed for saquinavir. Hydrophobicity was promoted by the incorporation of cholesterol, hence nevirapine loading was significantly increased but not alter that of saquinavir. An optimized encapsulation of 45% and 30% was achieved for nevirapine and saquinavir, respectively, at a combinatory ratio of 9:1:1 for EPC (1,2 Dioleoyl sn-glycero 3 ethylphosphocholine): Cholesterol: distearoylphosphatidylethanolamine (DSPE)-PEG. The study also employed specific targeting, homing on HIV-infected T-cells and macrophages by surface modification of the liposomes with anti-CD4 antibodies. Overall, the anti-CD4 immuno-liposomes demonstrated enhanced uptake into the CD4-positive Jurkat cells, exhibited preferential delivery to HIV-infected cells, and the proliferation of the virus was markedly reduced by dual activity of the encapsulated drugs. These effects were at a lower concentration versus with the free drugs.

2.2 Tuberculosis

The *Mycobacterium tuberculosis* (MTB) is a pathogenic bacterium with inherently robust transmission and survival mechanism, causing one of the first and second deadliest infectious diseases —tuberculosis (TB) [46]. It mainly affects the lungs, but its target organs are vast, including the pleura, bones, joints, meninges, genitourinary tract, and the skin, challenging both diagnosis and therapeutic interventions [46]. Like HIV, it exploits various mechanisms to evade the hosts' immune system as well as exogenous drugs, wherein its escape from the lungs' first line of defense, the macrophages leads to the progression of the disease from the development of granulomas to cavity formation where unabated replication of MTB occurs. Significant challenges in the treatment of TB exist at different levels of disposition, including at the molecular, physiological, clinical, and geographical levels, which has led scientists to consider disease subgroupings for selecting appropriate disease biomarkers [47]. Hence, to date, TB remains a global dilemma, which was responsible for 1.3 million deaths in 2023 [48]. Although this is a 19% decline from 2015, drug resistance is a continuous concern with the development of multiple drug resistance (MDR) and extensively drug-resistant (XDR) TB —a challenge due to the lack of established clinical management systems in terms of specific/tailored diagnosis and curative drug treatments. This is a call for innovative methods of treatment, such as resistance to first-line and second-line drugs, and importantly, the scarcity in the approval of the new drugs: bedaquiline, delamanid, and pretomanid, which were discovered 40 years after the approval of the latest of the second generation of drugs [49, 50], are indications for improving the drug delivery system as opposed to discovery of new drugs.

Liposomes create a new dimension in solving the pharmacokinetic and pharmacodynamic associated with TB drugs, such as the long duration of treatment, side effects that are debilitating in physical and mental health with subsequent relapse rates and the development of drug resistance. By encapsulating TB drugs in liposomes, which can be designed to target specific cells, this reduces the chances of off-targeting; while improving concentration efficacies, multi-drugs can then be loaded and simultaneously delivered in a single dose. PEGylation (and other methods) can be used to increase the longevity and bioavailability of the drug, and encapsulation can facilitate the reduction in the premature metabolism of the drug. Consequently, lower doses and shortened periods of treatment will negate issues of patient compliance and decrease relapse rates, lowering the prospective of drug resistance.

Since TB is chiefly a disease of the lungs, the pulmonary and parenteral routes of administration are better qualified as they avoid rapid metabolic degradation and gastrointestinal absorption, which result in the delivery of suboptimal concentrations in the diseased alveoli. Aerosol diagnostics in the form of volatile organic compounds (VOC) and inhaled therapeutics have been in the sphere of TB research since the 1940s, promising an efficient route for theranostics and prevention of drug resistance [51]. These present as easier and less invasive methods of sampling and drug administration. Encapsulated drugs can be delivered in suspension or dry powder form by pulmonary devices in small doses using pressurized metered inhalers (pMDIs), dry powder inhalers (DPIs), and soft mist inhalers (SMIs) versus medical nebulizers, which deliver larger doses of medicine [23]. Liposomes present an ideal mode of transport due to their remarkable uptake by macrophages, biocompatible nature, and surfactant-like properties similar to the lungs [52], which lowers the chances of localized toxicity and adverse reactions. In this regard, a few interesting studies have been dedicated to investigating the purpose of old and new drugs for inhalation therapy [53].

In a study by Bhardwaj and colleagues, liposomes were surface modified with mannan, a macrophage-specific receptor that is associated with MTB binding and activation of the innate immune and loaded with the first-line antituberculosis drugs rifampicin (RIF), isoniazid (INH), and pyrazinamide (PYZ) [54]. The mannan-liposome encapsulating RIF, INH, and PYZ were lyophilized into an inhalable powder, which was cryopreserved using sucrose, an important component for preserving hydration of the phospholipid heads and preventing drug leakage. Due to their differing hydrophobicity indices, their localization was specific. Whereas, the lipophilic RIF intercalated with lipid bilayer, the more hydrophilic INH and PYZ were entrapped proximal the aqueous core. This may also be the cause for their drug release of 67.4 ± 2.43, 65.8 ± 2.56, and $58.1 \pm 1.65\%$ for RIF, INH, and PYZ, respectively. This was similar to the neutral liposomes and followed the Korsemeyer-Peppas model for drug release; in contrast, the lung uptake of the mannan-liposomes was higher, 51.36 and 48.66% for the neutral liposomes [54]. As it has been clear throughout the chapter, the lipid composition of liposomes can be varied to perform specific functions and dual functionality. This was demonstrated by Chimote and Banarjee, who successfully prepared DPPC-based liposomes entrapping INH, thus acting as anti-TB and anti-atelectatic surfactant. The authors exploited the well-known fact that DPPC is the most abundant lipid of the lung surfactant as well as an exogenous surfactant [55]. The surfactant plays an essential role in preventing the lung collapse that becomes apparent in the late stages of pulmonary (PTB), enhancing the anti-TB reach of INH; moreover, the drug release was sustained for over 24 hours. This is an insightful finding, which presents a strong case for designing liposomes that can target the different phenotypic stages of the disease such as the granulomatous, necrotic granulomatous, caseous necrotic granulomatous, and cavitary disease, especially where evasion is prominent.

The physicochemical properties of the microenvironment of the diseased tissue must be considered as factors such as pH can alter the structure and function of the liposome and are especially relevant for MTB infection, which is characterized by acidification steps during phagolysosomal fusion. A follow-up study by Bhardwaj and team, who designed a pH-sensitive mannan-anchored encapsulating INH and Ciprofloxacin HCl (CIP HCl), shows a drug release of 58-64% in alkaline pH compared to 81-87% release in macrophage pH [54]. Some of the prominent research into aerosol therapeutics for TB have been reviewed by Hickey et al., highlighting the advantages, progresses, and limitations in targeting infected alveoli by inhalers versus nebulizers, mostly advocating for the drug powder inhalers as being more efficient [51].

Macrophages can likewise be specifically targeted by tufstin-modified liposomes. The treatment of MTB-infected mice with tufstin liposomes loaded with RIF displayed an impressive 2000-fold higher reduction in bacterial load in the lungs compared to the free drug [56]. This was achieved without continuous administration of the drug, thus lowering toxicity that is associated with high-drug dosages —owing to specific targeting of macrophages.

The adaptability of liposomes is once again demonstrated by the thermosensitive liposome-in-hydrogel, which targets the bones [57]. Bone TB is a common EPTB form, accounting for 10% of the cases; as this side is inaccessible to drugs owing to poor blood supply, treatment includes surgical removal of the diseased tissue followed by localized injection of a multi-anti-TB drugs —a painful and invasive method of treatment. INH is the foremost drug of treatment for TB infecting the bones, but its hydrophilic properties predispose it to burst drug release and shortened periods of release, which were greatly overcome by using its hydrophobic derivative N′-Dodecan oylisonicotinohydrazide (DINH). Additionally, the DINH encapsulated in liposomes, followed by the hydrogel adds another network of protective layer, which sustained drug release over 24 hours compared to the 4 hours when only the hydrogel was used [57]. The thermosensitive peptide PLGA-PEG-PLGA with self-healing properties was an excellent incorporation into the liposomes, which meant that upon contact and sensing of bodily temperature the hydrogel transforms into a sol-gel, which is less painful than other implantations while inducing self-healing at the diseased site. Liposome-in-hydrogel seems an effective method for targeting the bone tissue, and adhesion to the bone tissue can be enhanced by appending functional groups, such as octadecylamine, which is an amphiphilic cationic film-forming material, and sulfhydryl groups [57]. These modified liposomes showed better bone regeneration and remodeling. EPTB can be effectively treated using liposome-encapsulated TB drugs, such as rifabutin (RFB), which is used to treat patients who are intolerant to rifampicin including HIV-coinfected individuals. Gaspar and colleagues show that DPPC: DPPG formulated liposomes act as more efficient delivery systems for RFB displaying high concentrations of RFB in the liver, spleen, and lung within a 24-hour period and decreased bacterial loads in these organs of a disseminated-TB rat model [58].

The introduction of asymmetric liposomal systems has broadened the design opportunities for drug and vaccine delivery, with some important breakthroughs such as the commercialized and FDA-approved liposome-based amikacin suspension Arikace® —the first and only approved inhalable drug for treating the nontuberculous *Mycobacterium avium* complex (MAC), previously difficult to treat [23, 59]. Asymmetry in membranes is a common phenomenon in eukaryotes, which was first described by Bretscher in 1972, where the distribution in the type of lipids within the inner and outer leaflets is deliberately organized differently. It is also noteworthy that the protein composition and cholesterol distribution also differs [23]. Phosphatidylcholine (PC) and sphingomyelin (SM) preferentially makeup the outer leaflet, while the negatively-charged lipids that include phosphatidylserine, phosphatidylethanolamine, and phosphatidylinositol are embedded in the inner leaflet [60].

This asymmetry is maintained *via* the flippases, floppases, and scramblases, performing intricate functions, such as the translocation of the lipids across the membranes [23, 60]. Moreover, other effectors of asymmetry exist, such as the size of the vesicle, which observable augments asymmetry with decreasing diameter size [23].

2.3 Malaria

Malaria is a vector-borne disease that is caused by the protozoan parasite, *Plasmodium*, transmitted by the female *Anopheles* species mosquitoes [3, 61]. The incidences of malaria span 85 countries, infecting 249 million people and culminating with 608,000 deaths, with 94-95% of these occurrences based in African regions [62]. There are five different species of *Plasmodium* that can cause pathogenesis, which is characterized by the well-established cycle between the mosquito and the human, presenting with similar events across the different species. Briefly, the sporozoites from the mosquito's salivary glands are injected into the human host *via* a bite, traveling *via* the bloodstream to the liver where asexual replication occurs. Following that, the thousands of daughter merozoites flood the bloodstream invading the red blood cells, and this is a crucial stage of adverse symptom manifestations and also a great target for drugs. A full circle is achieved upon another event of a mosquito bite, wherein the merozoites turned gametophytes in the erythrocytes are ingested into the salivary glands of the mosquito [3, 61].

The enormous effort toward developing antimalarial drugs has been greatly compromised by the persistent counter evolution of drug resistance in parasites, and this is resistance against almost every known drug for treating humans [63, 64]. Quite concerning is the spontaneity of the mutations occurring without any drug pressure [65]. This is within the existing shortcomings in the drug discovery pipeline of malaria drugs, including lack of specificity in the intracellular parasite-associated biomarker targets, poor drug solubility, low permeability, and poor bioavailability; consequently, high toxicity as a result of overcompensation by giving high doses becomes apparent, which also compounds issues of noncompliance in patients due to accompanying side effects [63, 64]. It is interesting that despite the dramatic reduction in malaria deaths owing to administration of quinoline-based drugs, their mode of action has not been fully deciphered albeit the 400 years following their discovery [66]. Emerging drug resistance has posed a serious bottleneck in the treatment of malaria, this is true for quinolone-based drugs chloroquine, mefloquine, and halofantrine as well as artemisinin, which is not in this group. Precautionary use of primaquine to avoid resistance and relapse has been a continuous discussion. Liposomes have proved distinct adeptness in mitigating these shortfalls among other interventions that exploit nanotechnology, with several studies reporting improved compound pharmacokinetics and pharmacodynamics from *in vivo* and *in vitro* studies of treatment with the liposome-encapsulated drug versus the drug on its own [64].

To elaborate on the role of electrostatic interactions in the structure-function relationship in optimizing encapsulation, the drug loading capacity of liposomes with platinum-chloroquine (PtCQ) diphosphate dichloride, a potent combination against chloroquine-resistant cells was investigated [67]. PtCQ was loaded into PEGylated liposomes that were either cationic or neutral. Since the inner membrane of erythrocytes is anionic, the exposure that occurs following eryptosis makes it more likely for cationic liposomes to bind, increasing the chances of drug delivery. Although this warrants using cationic liposomes, the challenge remains with the low half-life and high chance of macrophage recognition, hence a common mitigation was PEGylation of the liposomes [67]. Both PEGylated cationic and neutral liposomes were able to prevent leakage of the drug; moreover, the encapsulation efficiency was as high as 76.1% for neutral liposomes and 96.9% for the cationic liposomes. This outcome is not surprising, given the physicochemical relationship of the interacting liposomes and the erythrocytes.

PEGylation is also an alternative to using pH gradient methods, which maximize the loading capacity of small-sized liposomes that are characterized by low drug loading [68–70]. The pH gradient methods conveniently depend on the intrinsic structure of the transmembrane where a progressive drop in pH occurs toward the core of the liposome, and the efficiency of drug upload is dependent on its overall counter charge as well as other physicochemical properties [68, 69]. Knowledge of the ionization states of the drugs can shorten the time spent on optimizing the compatible conditions for encapsulation when predictive models are exploited; the modeling of the liposomal distribution of diprotic antimalarial compounds has been analyzed by Moles and co-workers [71]. The distribution of quinine, primaquine, tafenoquine, quinacrine, and chloroquine in phosphatidylcholine-based liposomes, as well as interest in the application of pH gradients as a method of efficient drug uploading and their responsiveness, were studied [71]. This gradient method is particularly suitable for addressing the non-endocytic uptake of drugs by cells, such as the erythrocytes, wherein their membrane skeleton is a meshwork of spectrin-actin molecules that exploit steric hindrance to inhibit endocytosis, mainly relying on the diffusion pathway for exchange of molecules [72]. With respect to NP, the inhibitory effect is due to membrane-NP interaction, size of NP relative to the meshwork, and skeleton tension as was found through molecular simulation models using dissipative particle dynamics (DPD). This further reiterates the structure-function relationship as a useful comprehension and building tool for expediting the development of predictive computational models that would facilitate the surface modification of liposomes, making them a more specific and efficacious system for drug delivery.

Ligands are convenient in that they can be used as structural moieties to modify biological macromolecules to serve specific functions or applications. Active pharmaceutical compounds that display target promiscuity, with non-validated biomarkers or non-elucidated mechanisms of action, can benefit from liganded liposomes. A wide spectra of ligands have been used to modify liposomes for targeting malaria-infected cells, ranging from derivatives of glycolipids and peptides to antibodies; these studies have been summarized in a review by Memvanga and Nkanga [64]. Varying lipid compositions and ratios have been explored for optimizing the efficiency of the liganded moieties.

Peptide-functionalization of nanomaterials, such as liposomes, is an ingenious practice that has been adopted for enhancing the bioavailability and specificity of small molecule delivery, in principle creating an active targeting system [73]. A 19-amino acid (19-aa)-long peptide sequence from the circumsporozoite protein (CSP) has been used for targeting the hepatic stage of the malaria virus by incorporating it into liposomes. CSP and the thrombospondin-related anonymous protein are highly specific to the hepatocytes, wherein their localization to the liver is observed within minutes of sporozoites being introduced to the bloodstream [74]. Subsequently, liposomes decorated with a 19-aa CSP sequence accumulated with more than several hundred-fold higher in the liver compared to other organs, except for the 10-fold higher accumulation than in the spleen; and the binding is suggested to be *via* the heparin-associated proteoglycans [74]. Since the macrophages in the spleen are known to clear particles, then this accumulation is warranted. This system indeed presents an efficient solution for preerythrocytic specific targeting of malaria; however, the antigenicity needs to be investigated prior to therapeutic use. Another peptide-based strategy was developed for targeting the intraerythrocytic stage, wherein a macrophage activating tetrapeptide, tuftstin, and mimicking a sequence in the Fc-portion of the heavy chain of the IgG [75] was used to surface modify liposomes. This concept stems from the ability of tuftstin to activate cells of the immune, such as macrophages, to kill intraerythrocytic malarial parasites. Pretreatment of mice with hydrophobic derivatives of tufstin incorporated into liposomes resulted in reduced mortality of *Plasmodium berghei* infected mice as well as decreased parasitemia [76]. These studies not only open novel avenues for developing prophylactic treatment for malaria but also for the development of vaccines.

Attaching antibodies to nanomaterials is a widely used preparation method that endows specificity to therapeutic and diagnostic products that can be used for liposome DDS. This is evident in the covalent conjugation of an anti-erythrocyte F(ab')2 and liposomes encapsulating chloroquine used in treating mice infected with *Plasmodium berghei* as well as chloroquine-resistant mice [77]. Studies showed that the F(ab') two-bearing antibodies enhanced binding to the erythrocytes, reducing parasitemia in susceptible and resistant mice, and prolonging survival [77]. This strategy presents a more straightforward way of increasing the sensitivity and specificity of liposomes.

3. Computational tools in the design of liposome-based drug delivery systems

To expedite the liposome technology, predictive computational models are proficient tools that can be employed by adopting high-throughput screening methods, which further reduce time spent and cost required in the drug discovery pipeline, and for developing optimal conditions *in vitro*. Leveraging existing empirical data and structure-function relationships, it is possible to create such models. Layers of complex algorithms and machine learning systems have afforded computational modeling with the ability to handle multiple variables simultaneously, better accuracy of predictions, data output that is simpler to analyze, and improved spatiotemporal landscapes — referred to as coarse grained analysis.

With the capabilities of various simulation methods, different analyses can be extracted, including from molecular dynamics (MD), machine learning (ML), Monte Carlo (MC), finite element analysis (FEA), computational fluid dynamics (CFD), density functional theory (DFT), and dissipative particle dynamics (DPD) [78]. At the molecular level, MD can be used to analyze the behavior of atoms with respect to space, time, and the forces acting on them, giving an overview of molecular interactions. In this case, a motion-based equation that is derived from Newton's theory of force enables the prediction of behavior of liposomes with the drug and the environment. With MC, statistical analysis and probability theory are exploited in simulation of complex systems, wherein various potential scenarios or outputs are analyzed. MC allows the prediction of drug release and uptake, bioavailability, and toxicity can be predicted [78].

In CFD modeling, the flow of fluids in cells and tissues are described; thus, the movement of DDS can be better comprehended using this technique, while the FEA model also exploits mathematical algorithms to facilitate the analysis of DDS under different conditions, thus testing the mechanical properties of the liposome [78].

The CFD model has been greatly used in the design of inhalers and nebulizers, by simulating air flow into the lungs and drug absorption — the design of metered-dose inhaler (MDI), also known as the AeroCup for COVID-19 being exemplary [79]. Another computational model that has been successfully used for DDS is DFT, wherein the thermodynamic nature of interacting systems can be studied, giving insight into the energetic changes that occur at the molecular level; this method is also capable of defining the geometric structure and electrical characteristics of nanocarriers upon drug docking [80].

Machine learning and other AI tools bring another complex element to the computational predictive modeling. By employing training data from known information based on the respective structure-function relationships of DDS or in unknown cases, the data is adopted from quantum models; these tools can facilitate the prediction and modeling of DDS in conditions that are controlled virtually, which would otherwise be impossible to replicate using empirical models [78, 80]. For instance, the stability/dispersity and size of the liposome were predicted using the support-vector machine models and feed-forward artificial neural networks, respectively —initially trained using over 200 liposomes that were varied in their composition, including flow rates, lipid concentrations, and organic: water ratios [81]. This work aimed to optimize the pharmacokinetics of curcumin-loaded liposomes based on the influence of their physicochemical properties and their interactions, which ultimately have an effect on biocompatibility, toxicity, and interaction with biological material [81]. These computational models are adaptable to other DDSs such as dendrimers, polymer-based nanoparticles, solid-lipid nanoparticles, and implantable DDS [78].

Simulations with the lipid bilayers have been carried out, describing the intra- and intermolecular interactions between components of the bilayer itself, and with the cargo as well as distribution of cargo within the liposome [15, 19, 71]. The length of acyl groups in phospholipids influences their packing geometry and conformation of the phospholipid [19], this is due to the type of dominating forces, either stabilizing or de-stabilizing affecting the liposomes' fluidity. Macromolecules are held together by various forces, including hydrophobic, hydrogen bonds, van der Waals, ionic, and/or electrostatic interactions. These forces can be manipulated to efficiently load specific cargo while controlling their rate of release, wherein the localization of the drug is dependent on its hydrophobicity index, either intercalating in the lipid bilayer or proximal to the aqueous core for a slower release. The ionization state of the cargo is equivocally integral to its retention and release from the liposome, making both the liposome and cellular microenvironments relevant to their efficiency. An example of bilayer simulations is illustrated in **Figure 3**.

Hydrogen bonds are agreeably one of the vastest macromolecular "glues" in existence, with their directionality highly sought as recognition elements in biomolecules, spanning enzymes, proteins, nucleic acids, lipids, and others. In liposomal formulations, the molecular simulations have been dedicated to studying hydrogen bonds between the phospholipids, cholesterol, and water molecules within the lipid bilayer. Although the existence of H-bonds between the OH group of cholesterol and the polar head group of phospholipids has been supported, Pandit and team went a step further, suggesting that the CH···O H-bonding occurs between the methyl group from phospholipid and the oxygen atom of the cholesterol group, providing the rationale for formations that are larger than ratios of 1:1 [19]. CH···O H-bonding has been debatable; however, quantum studies have

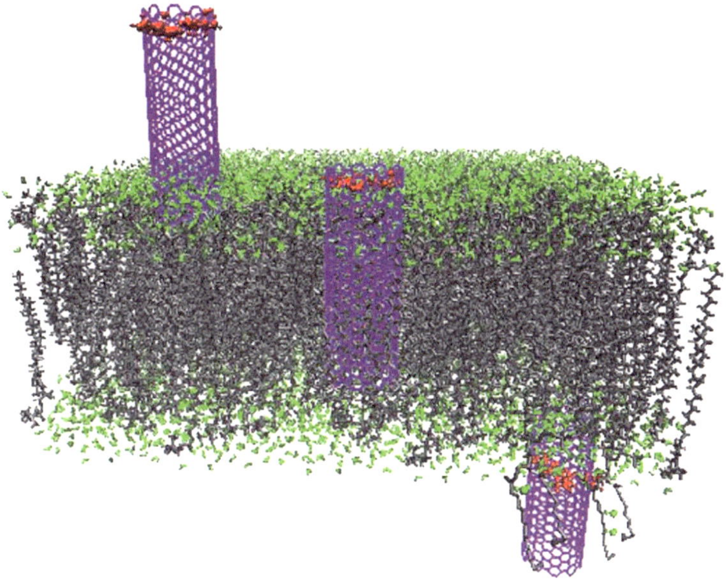

Figure 3.
Simulation of a single-walled nanotube loaded with doxorubicin docking into the bilayer membrane. (This figure is reproduced from a review by Salahshoori and colleagues [78, 82] with open access from the creative commons CC-BY license).

characterized them as weaker and with minute directionality than the conventional H-bonds [83]. In structural biology, these bonds were only appreciated about 40 years ago, with immense contributions to the structural integrity of biomolecules, recognition and binding, and catalysis [84, 85]. In conclusion, Pandit and colleagues found that two types of complexes exist in each mixture of dilauroylphosphatidylcholine (DLPC): cholesterol and (DPPC): cholesterol, exhibiting stoichiometries of 1:1 and 2:1; however, the DLPC: cholesterol was more populated with species of a 1:1 ratio and vice versa for the DDPC: cholesterol. This difference was attributed to the different acyl lengths of the phosphatidylcholine lipids. CH···O H-bonds were defined as crucial for the formation of the complexes and in their aggregation, especially in the DPPC: cholesterol [19]. These types of computational models are relevant for multidrug loading and their timeous release; hence, the composition of the liposome can be designed and refined *in silico* with simultaneous docking of drugs, such that the correct partitioning is achieved to obtain optimized pharmacokinetic and pharmacodynamic properties.

4. Conclusions

The role of liposomes in nanomedicine is futuristic, with the successful approval and commercialization of DaunoXome®, Doxil®, AmBisome®, Arikace®, RUTI®, and Mosquirix®, which are used in either treating or preventing HIV, TB, and malaria and/or associated diseases. The ongoing pipeline developments and clinical trials depict endless possibilities. Computational tools can facilitate the fast tracking of this technology, wherein optimization processes can be minimized in terms of time and cost as well as define the limitations.

Author details

Palesa Pamela Seele
Advanced Materials Division, Nanotechnology Innovation Center, Health Platform, Mintek, Randburg, South Africa

*Address all correspondence to: palesas@mintek.co.za

References

[1] Bulbake U, Doppalapudi S, Kommineni N, Khan W. Liposomal formulations in clinical use: An updated review. Pharmaceutics. 2017;**9**(2):12-45. DOI: 10.3390/pharmaceutics9020012

[2] He Y et al. Liposomes and liposome-like nanoparticles: From anti-fungal infection to the COVID-19 pandemic treatment. Asian Journal of Pharmaceutical Sciences. 2022;**17**(6):817-837. DOI: 10.1016/j.ajps.2022.11.002

[3] Suhaimi NAA et al. Application of liposomes in the treatment of infectious diseases. Life Sciences. 2022;**305**:120734-1207349. DOI: 10.1016/j.lfs.2022.120734

[4] Hedt BL, Laufer MK, Cohen T. Drug resistance surveillance in resource-poor settings: Current methods and considerations for TB, HIV, and malaria. American Journal of Tropical Medicine and Hygiene. 2011;**84**(2):192-199. DOI: 10.4269/ajtmh.2011.10-0363

[5] Alene KA et al. Spatial codistribution of HIV, tuberculosis and malaria in Ethiopia. BMJ Global Health. 2022;**7**(2):e007599-e007697. DOI: 10.1136/bmjgh-2021-007599

[6] Aslan Y, Atabay M, Chowdhury HK, Göktürk I, Saylan Y, Inci F. Aptamer-based point-of-care devices: Emerging technologies and integration of computational methods. Biosensors. 2023;**13**(5):569. DOI: 10.3390/BIOS13050569

[7] Trucillo P, Campardelli R, Reverchon E. Liposomes: From bangham to supercritical fluids. PRO. 2020;**8**(9):1022-1037. DOI: 10.3390/pr8091022

[8] Sharma VK, Agrawal MK. A historical perspective of liposomes-a bio nanomaterial. Materials Today: Proceedings. 2021;**45**:2963-2966. DOI: 10.1016/j.matpr.2020.11.952

[9] Rommasi F, Esfandiari N. Liposomal nanomedicine: Applications for drug delivery in cancer therapy. Nanoscale Research Letters. 2021;**16**(1):95-115. DOI: 10.1186/s11671-021-03553-8

[10] Siepmann J et al. Lipids and polymers in pharmaceutical technology: Lifelong companions. International Journal of Pharmaceutics. 2019;**558**:128-142. DOI: 10.1016/j.ijpharm.2018.12.080

[11] Robert J. Preclinical assessment of anthracycline cardiotoxicity in laboratory animals: Predictiveness and pitfalls. Cell Biology and Toxicology. 2007;**23**:27-37. DOI: 10.1007/s10565-006-0142-9

[12] Balazsovits JAE et al. Analysis of the effect of liposome encapsulation on the vesicant properties, acute and cardiac toxicities, and antitumor efficacy of doxorubicin. Cancer Hemotherapy and Harmacology. 1989;**23**:81-86

[13] Bally MB, Nayar R, Masin D, Hope MJ, Cullis PR, Mayer LD. Liposomes with entrapped doxorubicin exhibit extended blood residence times. Materials Today: Proceedings. 1990;**45**:2963-2966

[14] Bozzuto G, Molinari A. Liposomes as nanomedical devices. International Journal of Nanomedicine. 2015;**10**:975-999. DOI: 10.2147/IJN.S68861

[15] Wennberg CL, Van Der Spoel D, Hub JS. Large influence of cholesterol on solute partitioning into lipid membranes. Journal of the American Chemical Society. 2012;**134**(11):5351-5361. DOI: 10.1021/ja211929h

[16] Risselada HJ. Cholesterol: The plasma membrane's constituent that chooses sides. Biophysical Journal. 2019;**116**(12):2235-2236. DOI: 10.1016/j.bpj.2019.05.003

[17] Pinkwart K et al. Nanoscale dynamics of cholesterol in the cell membrane. Journal of Biological Chemistry. 2019;**294**(34):12599-12609. DOI: 10.1074/jbc.RA119.009683

[18] Yang J, An HW, Wang H. Self-assembled peptide drug delivery systems. ACS Applied Bio Materials. 2021;**4**(1):24-46. DOI: 10.1021/acsabm.0c00707

[19] Pandit SA, Bostick D, Berkowitz ML. Complexation of phosphatidylcholine lipids with cholesterol. Biophysical Journal. 2004;**86**(3):1345-1356. DOI: 10.1016/S0006-3495(04)74206-X

[20] Rowlands LJ, Marks A, Sanderson JM, Law RV. 17O NMR spectroscopy as a tool to study hydrogen bonding of cholesterol in lipid bilayers. Chemical Communications. 2020;**56**(92):14499-14502. DOI: 10.1039/d0cc05466f

[21] Presti FT, Pace JRJ, Chan SI. Cholesterol-phospholipid interaction in membranes. 2. Stoichiometry and molecular packing of cholesterol-rich domains?. Biochemistry. 1982;**21**:3831-3835

[22] Bhattacharya S, Haldar S. Interactions between cholesterol and lipids in bilayer membranes. Role of lipid headgroup and hydrocarbon chain-backbone linkage. [Online]. Biochimica et Biophysica Acta. 2000;**1467**:39-53. Available from: www.elsevier.com/locate/bba

[23] Al Badri YN, Chaw CS, Elkordy AA. Insights into asymmetric liposomes as a potential intervention for drug delivery including pulmonary nanotherapeutics. Pharmaceutics. 2023;**15**(1):294-316. DOI: 10.3390/pharmaceutics15010294

[24] Danaei M et al. Impact of particle size and polydispersity index on the clinical applications of lipidic nanocarrier systems. Pharmaceutics. 2018;**10**(2):57-74. DOI: 10.3390/pharmaceutics10020057

[25] Labiris NR, Dolovich MB. Pulmonary drug delivery. Part I: Physiological factors affecting therapeutic effectiveness of aerosolized medications. British Journal of Clinical Pharmacology. 2003;**56**(6):588-599. DOI: 10.1046/j.1365-2125.2003.01892.x

[26] Bertrand N, Leroux JC. The journey of a drug-carrier in the body: An anatomo-physiological perspective. Journal of Controlled Release. 2012;**161**(2):152-163. DOI: 10.1016/j.jconrel.2011.09.098

[27] Gonzalez Gomez A, Hosseinidoust Z. Liposomes for antibiotic encapsulation and delivery. ACS Infectious Diseases. 2020;**6**(5):896-908. DOI: 10.1021/acsinfecdis.9b00357

[28] Németh Z et al. Quality by design-driven zeta potential optimisation study of liposomes with charge imparting membrane additives. Pharmaceutics. 2022;**14**(9):1798-1823. DOI: 10.3390/pharmaceutics14091798

[29] Jarzynska K, Gajewicz-Skretna A, Ciura K, Puzyn T. Predicting zeta potential of liposomes from their structure: A nano-QSPR model for DOPE, DC-Chol, DOTAP, and EPC formulations. Computational and Structural Biotechnology Journal. 2024;**25**:3-8. DOI: 10.1016/j.csbj.2024.01.012

[30] Peetla C, Stine A, Labhasetwar V. Biophysical interactions with model lipid membranes: Applications in drug

discovery and drug delivery. Molecular Pharmaceutics. 2009;**6**:1264-1276. DOI: 10.1021/mp9000662

[31] Lou J et al. Advances in Oral drug delivery systems: Challenges and opportunities. Pharmaceutics. 2023;**15**(2):484-506. DOI: 10.3390/pharmaceutics15020484

[32] Al Ragib A, Chakma R, Dewan K, Islam T, Kormoker T, Idris AM. Current advanced drug delivery systems: Challenges and potentialities. Journal of Drug Delivery Science and Technology. 2022;**76**:103727-103747. DOI: 10.1016/j.jddst.2022.103727

[33] Ezike TC et al. Advances in drug delivery systems, challenges and future directions. Heliyon. 2023;**9**(6):e17488-17505. DOI: 10.1016/j.heliyon.2023.e17488

[34] Llor C, Bjerrum L. Antimicrobial resistance: Risk associated with antibiotic overuse and initiatives to reduce the problem. Therapeutic Advances in Drug Safety. 2014;**5**(6):229-241. DOI: 10.1177/2042098614554919

[35] Mitchell MJ, Billingsley MM, Haley RM, Wechsler ME, Peppas NA, Langer R. Engineering precision nanoparticles for drug delivery. Nature Reviews Drug Discovery. 2021;**20**(2):101-124. DOI: 10.1038/s41573-020-0090-8

[36] Chopra S, Venkatesan N, Betageri GV. Liposomes as nanocarriers for anti-HIV therapy. Drug Delivery and Translational Research. 2013;**3**(5):471-478. DOI: 10.1007/s13346-013-0134-2

[37] Bule M, Khan F, Niaz K. Antivirals: Past, present and future. In: Recent Advances in Animal Virology. © Springer Nature Singapore Pte Ltd.; 2019. pp. 425-446. DOI: 10.1007/978-981-13-9073-9_22

[38] Clercq E. Three decades of antiviral drugs. Nature Reviews Drug Discovery. 2007;**6**:941. DOI: 10.1038/nrd2485

[39] Back D, Marzolini C. The challenge of HIV treatment in an era of polypharmacy. Journal of the International AIDS Society. Feb 2020;**23**(2):e25449-51. DOI: 10.1002/jia2.25449

[40] Ramana LN, Sharma S, Sethuraman S, Ranga U, Krishnan UM. Investigation on the stability of saquinavir loaded liposomes: Implication on stealth, release characteristics and cytotoxicity. International Journal of Pharmaceutics. 2012;**431**(1-2):120-129. DOI: 10.1016/j.ijpharm.2012.04.054

[41] Phillips NC, Tsoukas C. Liposomal encapsulation of azidothymidine results in decreased hematopoietic toxicity and enhanced activity against murine acquired immunodeficiency syndrome. Blood. 1992;**79**:1137-1143

[42] Mallipeddi R, Rohan LC. Progress in antiretroviral drug delivery using nanotechnology. International Journal of Nanomedicine. 2010;**5**:533-547. [Online]. Available from: www.dovepress.com

[43] Jin SX, et al. Pharmacokinetics and tissue distribution of zidovudine in rats following intravenous administration of zidovudine myristate loaded liposomes

[44] Karunakaran B et al. Emerging trends in lipid-based vaccine delivery: A special focus on developmental strategies, fabrication methods, and applications. Vaccine. 2023;**11**(3):661-692. DOI: 10.3390/vaccines11030661

[45] Ramana LN, Sharma S, Sethuraman S, Ranga U, Krishnan UM. Stealth anti-CD4 conjugated immunoliposomes with dual antiretroviral drugs - Modern Trojan horses to combat HIV. European Journal

of Pharmaceutics and Biopharmaceutics. 2015;**89**:300-311. DOI: 10.1016/j.ejpb.2014.11.021

[46] Nair A et al. Advanced drug delivery and therapeutic strategies for tuberculosis treatment. Journal of Nanobiotechnology. 2023;**21**(1):414-445. DOI: 10.1186/s12951-023-02156-y

[47] Petruccioli E et al. Correlates of tuberculosis risk: Predictive biomarkers for progression to active tuberculosis. European Respiratory Journal. 2016;**48**(6):1751-1763. DOI: 10.1183/13993003.01012-2016

[48] WHO. Global Tuberculosis Report [online]. World Health Organization; 2023. Available from: https://www.who.int/news-room/fact-sheets/detail/tuberculosis [Accessed: June 03, 2024]

[49] Cohen KA, Manson AL, Desjardins CA, Abeel T, Earl AM. Deciphering drug resistance in mycobacterium tuberculosis using whole-genome sequencing: Progress, promise, and challenges. Genome Medicine. 2019;**11**(1):1-18. DOI: 10.1186/s13073-019-0660-8

[50] Liu Y et al. Delamanid: From discovery to its use for pulmonary multidrug-resistant tuberculosis (MDR-TB). Tuberculosis. 2018;**111**:20-30. DOI: 10.1016/j.tube.2018.04.008

[51] Hickey AJ, Durham PG, Dharmadhikari A, Nardell EA. Inhaled drug treatment for tuberculosis: Past progress and future prospects. Journal of Controlled Release. 2016;**240**:127-134. DOI: 10.1016/j.jconrel.2015.11.018

[52] Buya AB et al. Application of lipid-based nanocarriers for antitubercular drug delivery: A review. Pharmaceutics. 2021;**13**(12):2041-2065. DOI: 10.3390/pharmaceutics13122041

[53] Patil J, Devi V, Devi K, Sarasija S. A novel approach for lung delivery of rifampicin-loaded liposomes in dry powder form for the treatment of tuberculosis. Lung India. 2015;**32**(4):331-338. DOI: 10.4103/0970-2113.159559

[54] Bhardwaj A, Kumar L, Narang RK, Murthy RS. Development and characterization of ligand-appended liposomes for multiple drug therapy for pulmonary tuberculosis. Artificial Cells, Nanomedicine, and Biotechnology. 2013;**41**(1):52-59. DOI: 10.3109/10731199.2012.702316

[55] Chimote G, Banerjee R. In vitro evaluation of inhalable isoniazid-loaded surfactant liposomes as an adjunct therapy in pulmonary tuberculosis. Journal of Biomedical Materials Research. Part B, Applied Biomaterials. 2010;**94**(1):1-10. DOI: 10.1002/jbm.b.31608

[56] Agarwal A, Kandpal H, Gupta HP, Singh NB, Gupta CM. Tuftsin-bearing liposomes as Rifampin vehicles in treatment of tuberculosis in Micet. Antimicrobial Agents and Chemotherapy. 1994;**38**:588-593

[57] Liu P, Guo B, Wang S, Ding J, Zhou W. A thermo-responsive and self-healing liposome-in-hydrogel system as an antitubercular drug carrier for localized bone tuberculosis therapy. International Journal of Pharmaceutics. 2019;**558**:101-109. DOI: 10.1016/j.ijpharm.2018.12.083

[58] Gaspar MM et al. Rifabutin encapsulated in liposomes exhibits increased therapeutic activity in a model of disseminated tuberculosis. International Journal of Antimicrobial Agents. 2008;**31**(1):37-45. DOI: 10.1016/j.ijantimicag.2007.08.008

[59] Li Z et al. Characterization of nebulized liposomal amikacin

(Arikace™) as a function of droplet size. Journal of Aerosol Medicine and Pulmonary Drug Delivery. 2008;**21**(3):245-253. DOI: 10.1089/jamp.2008.0686

[60] Marquardt D, Geier B, Pabst G. Asymmetric lipid membranes: Towards more realistic model systems. Membranes. 2015;**5**(2):180-196. DOI: 10.3390/membranes5020180

[61] Cox FEG. History of the discovery of the malaria parasites and their vectors. Parasites and Vectors. 2010;**3**:1-9

[62] World Health Organization. World Malaria Report 2023 [online]. www.who.int; 2023. Available from: https://www.who.int/publications/i/item/9789240086173 [Accessed: June 03, 2024]

[63] Kamiya T, Paton DG, Catteruccia F, Reece SE. Targeting malaria parasites inside mosquitoes: ecoevolutionary consequences. Trends in Parasitology. 2022;**38**(12):1031-1040. DOI: 10.1016/j.pt.2022.09.004

[64] Memvanga PB, Nkanga CI. Liposomes for malaria management: The evolution from 1980 to 2020. Malaria Journal. 2021;**20**(1):327-360. DOI: 10.1186/s12936-021-03858-0

[65] Sinha S, Medhi B, Sehgal R. Challenges of drug-resistant malaria. Parasite. 2014;**21**:61-76. DOI: 10.1051/parasite/2014059

[66] Coban C. The host targeting effect of chloroquine in malaria. Current Opinion in Immunology. 2020;**66**:98-107. DOI: 10.1016/j.coi.2020.07.005

[67] Ibrahim S, Tagami T, Ozeki T. Effective-loading of platinum-chloroquine into pegylated neutral and cationic liposomes as a drug delivery system for resistant malaria parasites. Biological and Pharmaceutical Bulletin Journal. 2017;**40**(6):815-823. DOI: 10.1248/bpb.b16-00914

[68] Madden TD et al. The accumulation of drugs within large unilamellar vesicles exhibiting a proton gradient: A survey. Chemistry and Physics of Lipid. 1990;**53**:37-46

[69] Stensrud G, Sande SA, Kristensen S, Smistad G. Formulation and characterisation of primaquine loaded liposomes prepared by a pH gradient using experimental design. 2000. [Online]. Available from: www.elsevier.com/locate/ijpharm

[70] Qiu L, Jing N, Jin Y. Preparation and in vitro evaluation of liposomal chloroquine diphosphate loaded by a transmembrane pH-gradient method. International Journal of Pharmaceutics. 2008;**361**(1-2):56-63. DOI: 10.1016/j.ijpharm.2008.05.010

[71] Moles E, Kavallaris M, Fernàndez-Busquets X. Modeling the distribution of diprotic basic drugs in liposomal systems: Perspectives on malaria nanotherapy. Frontiers in Pharmacology. 2019;**10**(SEP):1064-1081. DOI: 10.3389/fphar.2019.01064

[72] Gao X, Yue T, Tian F, Liu Z, Zhang X. Erythrocyte membrane skeleton inhibits nanoparticle endocytosis. AIP Advances. 2017;**7**(6):065303-065315. DOI: 10.1063/1.4985052

[73] Liu M, Fang X, Yang Y, Wang C. Peptide-enabled targeted delivery systems for therapeutic applications. Frontiers in Bioengineering and Biotechnology. 2021;**9**:171504-171419. DOI: 10.3389/fbioe.2021.701504

[74] Longmuir KJ, Robertson RT, Haynes SM, Baratta JL, Waring AJ.

Effective targeting of liposomes to liver and hepatocytes in vivo by incorporation of a plasmodium amino acid sequence. Pharmaceutical Research. 2006;**23**(4):759-769. DOI: 10.1007/s11095-006-9609-x

[75] Agrawal AK, Gupta CM. Tuftsin-bearing liposomes in treatment of macrophage-based infections. 2000. [Online]. Available from: www.elsevier.com/locate/drugdeliv

[76] Gupta CM, Puri A, Jain RK, Bali A, Anand N. Protection of mice against plasmodium berghei infection by a tuftsin derivative. FEBS Letters. 1986;**205**(2):351-354. DOI: 10.1016/0014-5793(86)80927-9

[77] Chandra S, Agrawal AK, Gupta CM. Chloroquine delivery to erythrocytes in plasmodium Berghei-infected mice using antibody-bearing liposomes as drug vehicles. Journal of Biosciences. 1991;**16**:137-144

[78] Salahshoori I et al. Simulation-based approaches for drug delivery systems: Navigating advancements, opportunities, and challenges. Journal of Molecular Liquids. 2024;**395**:123888-123945. DOI: 10.1016/j.molliq.2023.123888

[79] Faqhrurrazi Abd Rahman M, Asmuin NZ, Taib I, Nazar J, Ismail A, Khairulfuaad R. Investigate flow characteristics of metered-dose inhaler (MDI) disposable inhaler spacer (aerocup) for COVID-19 patient by using computational fluid dynamic (CFD). CFD Letters. 2020;**12**(12):63-74. DOI: 10.37934/cfdl.12.12.6374

[80] Adekoya OC, Adekoya GJ, Sadiku ER, Hamam Y, Ray SS. Application of DFT calculations in designing polymer-based drug delivery systems: An overview. Pharmaceutics. 2022;**14**(9):1972-1992. DOI: 10.3390/pharmaceutics14091972

[81] Di Francesco V, Boso DP, Moore TL, Schrefler BA, Decuzzi P. Machine learning instructed microfluidic synthesis of curcumin-loaded liposomes. Biomedical Microdevices. 2023;**25**(3):29-42. DOI: 10.1007/s10544-023-00671-1

[82] Pakdel M, Raissi H, Shahabi M. Predicting doxorubicin drug delivery by single-walled carbon nanotube through cell membrane in the absence and presence of nicotine molecules: A molecular dynamics simulation study. Journal of Biomolecular Structure & Dynamics. 2020;**38**(5):1488-1498. DOI: 10.1080/07391102.2019.1611474

[83] Gu Y, Kar T, Scheiner S. Fundamental properties of the CH ··· O interaction: Is it a true hydrogen bond? Journal of the American Chemical Society. 1999;**121**(40):9411-9422. DOI: 10.1021/ja991795g

[84] Horowitz S, Trievel RC. Carbon-oxygen hydrogen bonding in biological structure and function. Journal of Biological Chemistry. 2012;**287**(50):41576-41582. DOI: 10.1074/jbc.R112.418574

[85] Derewenda ZS. C-H groups as donors in hydrogen bonds: A historical overview and occurrence in proteins and nucleic acids. International Journal of Molecular Sciences. 2023;**24**(17):13165-13190. DOI: 10.3390/ijms241713165

Section 4

Liposomes for Nutrient Delivery

Chapter 6

LiposoMax™ Liposomal PureWay-C®: A Liposomal-Vitamin C with Enhanced Biological Activity and Absorption

Benjamin S. Weeks and Pedro P. Perez

Abstract

Vitamin C is a water soluble nutrient that is required for human health and found in a wide range of foods. However, vitamin C must cross several tissue and cellular barriers to arrive at target tissues and intracellular sites. The barriers to vitamin C distribution in the body include the plasma membranes of the cells in the gut and vascular endothelium and also of the target cell and even subcellular compartments within the target cell. Due to the hydrophilic nature of vitamin C, facilitative transport proteins know as sodium-dependant vitamin C transporters –1 and –2 (SCVT-1 and SCVT-2) and glucose transporters (GLUTs) are required for this vitamin to cross the hydrophobic fatty acid tails of the phospholipid bilayer. However, more lipid soluble forms of vitamin C can also enter cells through diffusion and through this mechanism can enhance the rate and efficiency of vitamin C absorption from the gut and distribution throughout the body. For example, Ester-C® and PureWay-C® are both more lipid soluble forms of vitamin C and both have been shown to be more readily absorbed and distributed in the body that ascorbic acid. Indeed, PureWay-C® a phospholipid-associated lipid extracted form of vitamin C, has been shown to be best absorbed and the most bioactive form of vitamin C prior to liposomal forms. Currently, liposomal forms of vitamin C such as Lypo-spheric™ (LivOn Labs) and liposomal-PureWay-C® generated by LiposoMax™ technology (One Innovation Labs) have been shown to be better absorbed into cells. Further, liposomal-PureWay-C® has been shown to have enhanced capacity to support wound healing and neuronal cell function. This chapter presents the literature on the improved bioactivity and absorption of vitamin C when presented in diffusible lipid-extracted and liposomal forms.

Keywords: liposome, vitamin C, absorption, ascorbic acid, PureWay-C®, calcium ascorbate-calcium threonate-dehydroascorbate vitamin C, LiposoMax™, liposomal-PureWay-C®, absorption

1. Introduction

Liposomes are spherical phospholipid bilayers which have aqueous interior compartments capable of carrying hydrophilic molecules [1]. Liposomes have been used for a variety of purposes *in vitro* and *in vivo* to deliver a wide range of molecules to target cells including plasmids, cosmetics, pharmaceuticals and nutraceuticals [2–4]. Liposomes are able to fuse with the plasma membrane and deliver hydrophilic liposomal contents to the cell cytoplasm [1–4]. While drugs and nutrients may be able to cross membranes via facilitated transport, such as the SCVTs and GLUTs for vitamin C, liposome fusion with plasma membranes provides a more rapid diffusion of the liposomal contents to the cytoplasm [1, 2, 5, 6]. Further, incorporation of aptomers in the surface of the liposome for cell specific ligands has been shown to enhance cell and tissue targeting and tissue-specific uptake of liposomal contents [7–10]. For example, liposomes containing anticancer medication can be targeted to tumor tissues by incorporation of aptomers specific to the tumor onto the surface of the liposome lamellae [11] and such targeting may also be useful for vaccine development [7] and gene therapy [8–10] as well as in vitro research into cellular functions [12]. In addition to medicine, vaccines and gene therapies, enhanced nutrient uptake using liposomes has also been an important and aggressive area of research and development [13, 14]. This is particularly true for the hydrophilic nutrient, vitamin C, which has been shown to be better absorbed and provide enhanced bioactivity when incorporated into liposomes [15–22].

Vitamin C is an essential water soluble nutrient that serves as an enzyme cofactor in epithelia and neurons, as a signal transduction molecule in cells of the immune and nervous systems and as is an antioxidant that reduces the risk of cancer [5, 23, 24]. For example, vitamin C supplementation acts as a signal transduction molecule in leukocytes and functions to infections including sepsis, pneumonia and CoVid-19 and promotes T-cell maturation, neutrophil phagocytic activity, and cytokine and interferon production [25–36]. In neurons, vitamin C is an important cofactor for enzymes in dopamine synthesis and can signal calcium release and MAPK activity [37] and can stimulate neurite formation in hippocampal cell cultures [37]. Further, in human subjects, vitamin C supplementation protects patients from neurotoxins and reduces anxiety and depression and potentially neurodegenerative diseases [37]. Historically, vitamin C is known to be an important cofactor for collagen synthesis, skin health and wound healing [38–46]. Vitamin C also benefits the cardiovascular system, the liver and has anti-aging properties and it is this wide range of benefits from enhanced vitamin C absorption and distribution that has driven research into the development of liposomal and better absorbed forms of vitamin C [15–22].

Cellular absorption of vitamin C is regulated by transmembrane channel proteins known as sodium-dependant vitamin C transporters (SVCTs) and the glucose transporters (GLUTs) [5]. SCVTs are distributed in tissue-specific subtypes which allow different concentrations of vitamin C into the cells of different tissues [5]. For example, liver cells have intracellular vitamin C levels of 1 mM, while the brain and activated leukocytes can contain up to ten times this amount [5, 6]. However, rather than flooding channel proteins, improved gastrointestinal absorption and distribution of vitamin C has best been accomplished by increasing the lipid diffusible forms of vitamin C [16–19]. For example, when taken orally Ester-C® is better absorbed in human leukocytes compared to ascorbic acid [47]. Further, when compared to Ester-C®, a lipid metabolite of vitamin C with fatty acids (PureWay-C®) is better absorbed in human subject plasma and T-cells [48–50]. Further, PureWay-C® shows enhanced bioactivity including the stimulation of neurite outgrowth, epithelial cell attachment and immune cell function [48]. More recent work has shown that when packaged into liposomes, vitamin C is better absorbed into cells [16–19] and demonstrates enhanced bioactivity in wound healing assays [16]. Further, compared to nonliposomal-vitamin C, oral administration of liposomal-vitamin C leads to higher blood plasma concentrations of vitamin C and reduced risk of lipid peroxidation, hypertension and ischemia [17–20, 50]. Lastly, using the LiposoMax™ technology, which has been successfully used to incorporate a wide range of nutrients into liposomes (One Innovation Labs), liposomal-PureWay-C® has been shown to be better absorbed in human epithelial cells than nonliposomal-vitamin C and to enhance wound healing better than nonliposomal-vitamin C [16].

In addition to vitamin C, liposomal encapsulation has also been shown to enhance the absorption of other nutrients such as folate [51, 52], vitamin A [53, 54], vitamin D [55, 56], vitamin E [22, 57, 58], calcium and minerals [59, 60] among other nutrients [21, 53, 61–63]. However, of these supplements enhanced absorption in human subjects has only been done with in one study each with vitamin D3 [55], multivitamins [21] and minerals [60]. Incorporation of nutrients in general into liposomes is of great interest to the nutraceutical and health industries and that technologies such as LiposoMax™ (One Innovation Labs) are designed to deliver the future of liposomal nutrient packaging. However, this chapter focuses on liposomal-vitamin C since it is by far the most studied and prepared liposomal nutrient [13–22].

2. Bioactivity of lipid extracted vitamin C with fatty acids (PureWay-C®) and calcium ascorbate-calcium threonate-dehydroascorbate vitamin C

The biological activity of vitamin C is limited by the ability of this water soluble nutrient to arrive at target tissues and cell [5, 6]. In an effort to enhance vitamin C absorption beyond that seen with ascorbic acid and calcium ascorbate, more lipid soluble forms of vitamin C have been formulated and tested including an calcium ascorbate-calcium threonate-dehydroascorbate vitamin C (Ester-C®) and a lipid extracted vitamin C with fatty acids (PureWay-C®). These vitamin C preparations have been compared in the literature for a variety of biological activities including neurite formation (neuroplasticity), epithelial cell adhesion (wound healing), anti-inflammatory effects and antioxidant activity [48–50]. In all cases tested the more lipid soluble forms of vitamin C, calcium ascorbate-calcium threonate-dehydroascorbate vitamin C and lipid extracted vitamin C with fatty acids showed enhanced biological activity [48–50]. For example, vitamin C as ascorbic acid is known to enhance neurite outgrowth and support cellular microtubule-associate protein kinase (MAPK) cascade activity in PC12 cells [48]. When the various forms of vitamin C are compared, lipid extracted vitamin C with fatty acids showed a three-fold improved neurite outgrowth promoting activity compared to calcium ascorbate-calcium threonate-dehydroascorbate vitamin C [reviewed in [48–50]]. Further, lipid extracted vitamin C with fatty acids also reduces xenobiotic induced inflammatory markers in T-cells two-fold more than calcium ascorbate-calcium threonate-dehydroascorbate vitamin C and lipid extracted vitamin C with fatty acids have also been compared to the ability of ascorbic acid to reduce the oxidant damage and associated inflammation [48–50]. In specific, oxidative damage to tissues and the resultant cell-cell adhesion mechanisms in the inflammatory response to xenobiotics has been shown to be ameliorated best by lipid extracted vitamin C with fatty acids [48–50].

Again it is observed lipid extracted vitamin C with fatty acids consistently reduces xenobiotic induced inflammatory responses to a greater extent in human leukocytes compared to other forms of vitamin C [48–50]. Another biological activity requiring vitamin C uptake are some of the cellular events that underlie wound healing such as epithelial cell adhesion to fibronectin [5, 16, 23]. Vitamin C has been shown to promote fibroblast adhesion and wound healing properties and activity [5, 16, 23]. Indeed, lipid extracted vitamin C with fatty acids promotes enhanced fibroblast adhesion to and spreading on fibronectin to a greater extent than other vitamin C formulations tested in the literature [48]. Lastly, lipid extracted vitamin C with fatty acids shows more antioxidant activity than calcium ascorbate-calcium threonate-dehydroascorbate vitamin C, and while this activity may be relevant outside of cellular compartments to some extent, uptake of vitamin C from the gut is still required for the full antioxidant protective effect of vitamin C on tissues [49]. The idea that the fatty acid containing lipid extracted form of vitamin C is more biologically active due to improved uptake and absorption has been investigated as well and the topic of the next section.

3. T-cell absorption and uptake of lipid extracted vitamin C with fatty acids (PureWay-C®) in volunteer subjects

The enhanced biological activity of lipid extracted vitamin C with fatty acids has been shown to be due to the increased cellular uptake of this form of vitamin C in human T-cell when compared to ascorbic acid or calcium ascorbate-calcium threonate-dehydroascorbate vitamin C [48–50]. As noted above, vitamin C is an important nutrient in healthy immune system function [5, 16, 23, 24]. In order for vitamin C to exert these activities it must be taken-up and absorbed into the cell to a greater extent and or more quickly. Indeed, lipid extracted vitamin C with fatty acids is taken-up and absorbed into T-cells more rapidly than calcium ascorbate-calcium threonate-dehydroascorbate vitamin C [49, 50]. Calcium ascorbate-calcium threonate-dehydroascorbate vitamin C uptake in human T-cells reaches a maximum of approximately 30 nmol/mg cellular while lipid extracted vitamin C with fatty acids reaches a maximum of slightly over at 50 nmol/mg cellular protein [49]. The more rapid cellular localization of PureWay-C likely explains the enhanced immunological protection from xenobiotics afforded by the lipid extracted vitamin C with fatty acids formulation to T-cells [48, 49].

In addition to being better absorbed into human T-cells, lipid extracted vitamin C with fatty acids is also better absorbed into the human plasma after oral supplementation and results in higher serum vitamin C levels and reduced circulating levels of oxidized LDLs and C-reactive protein when compared to ascorbic acid, calcium ascorbate, and calcium ascorbate-calcium threonate-dehydroascorbate vitamin C supplementation [50]. Lipid extracted vitamin C with fatty acids is statistically significantly better absorbed (30% improvement) by the human body than calcium ascorbate-calcium threonate-dehydroascorbate vitamin C within the first 4 hours of supplementation [50]. Further, lipid extracted vitamin C with fatty acids supplementation also leads to the greatest reduction in circulating C-reactive protein and improvement in circulating oxidized LDLs in supplemented subjects [50]. Therefore, lipid extracted vitamin C with fatty acids is the best nonliposomal-vitamin C formulation for supplementation with regard to cellular uptake and absorption into the human body and biological activity.

4. Liposomal-PureWay-C®

While lipid-metabolite extraction had proved to provide the better absorbed vitamin C in the form of PureWay-C® [48], recent research has found that packaging vitamin C into liposomes provides even better absorption than calcium ascorbate-calcium threonate-dehydroascorbate vitamin C and PureWay-C® [16–20]. Compared to nonliposomal-vitamin C, oral administration of liposomal-vitamin C leads to higher blood plasma concentrations of vitamin C and reduced risk of lipid peroxidation, hypertension and ischemia [16–20]. Using the LiposoMax™ technology, vitamin C lipid metabolites have been incorporated into liposomes (liposomal-PureWay-C®) and is better absorbed in human epithelial cells than nonliposomal-vitamin C and to enhances wound healing above that seen with nonliposomal-vitamin C [16]. Therefore liposomal-PureWay-C® shows enhanced cellular distribution and biological activity [16]. **Figure 1** shows an electron micrograph of liposomal-PureWay-C® generated through the LiposoMax™ process. As noted, LiposoMax™ generated liposomal PureWay-C® shows enhanced absorption, wound healing capacity and

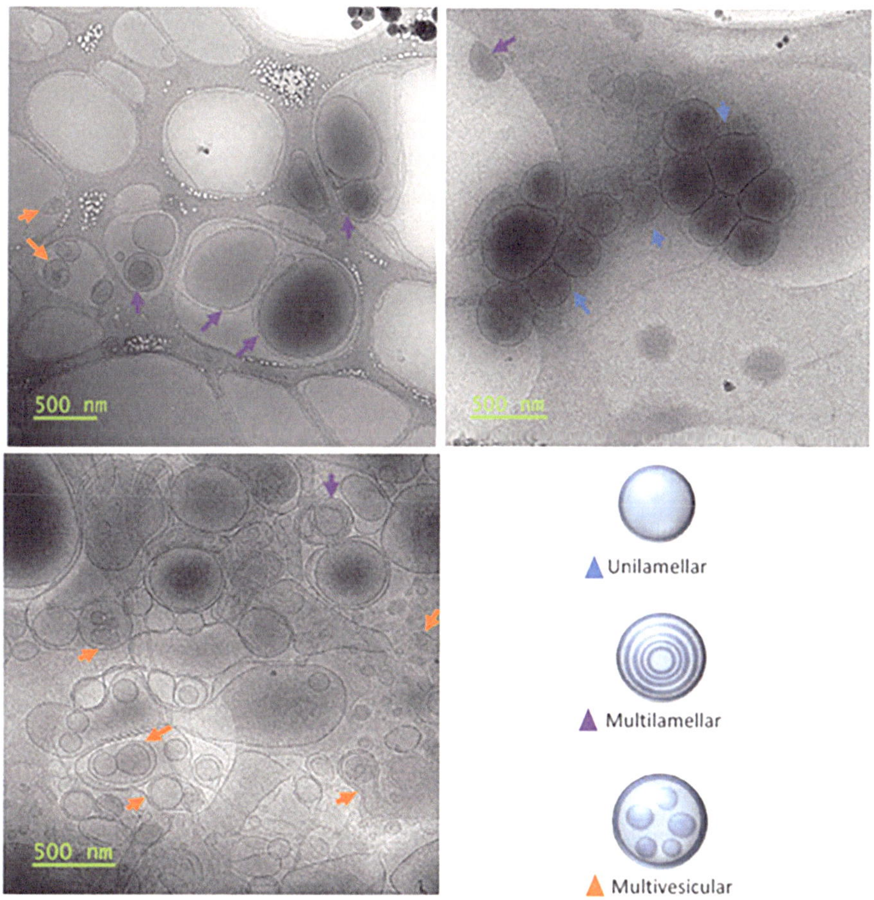

Figure 1.
Cryo-transmission micrographs of liposomal-PureWay-C®. All three panels are liposomal-PureWay-C® with liposomal size as small as 100 nm and unilamellar (blue arrows) multilamellar (purple arrows) and multivesicular (orange arrows) liposomes.

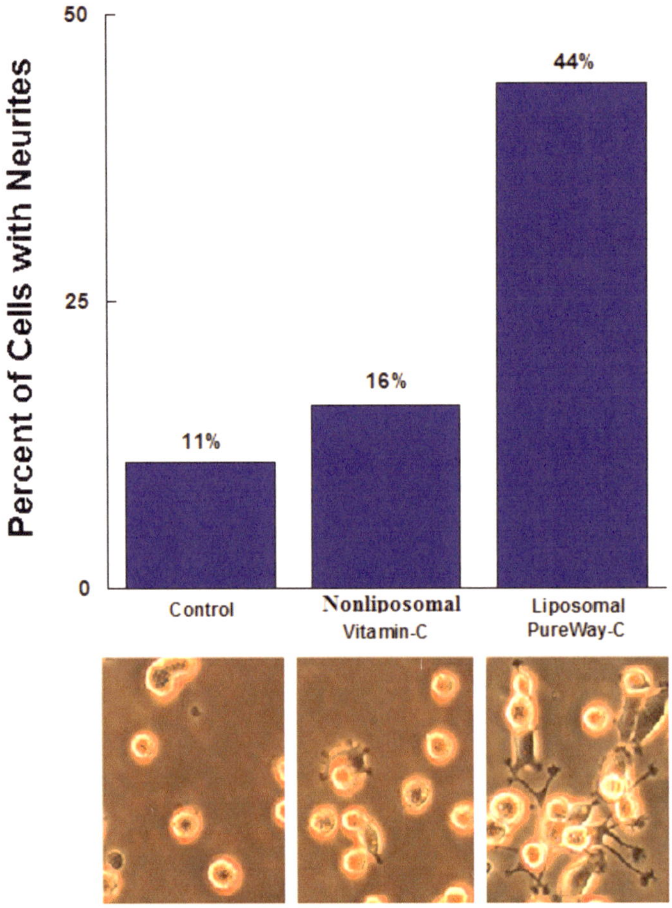

Figure 2.
Liposomal-PureWay-C® enhances neurite outgrwoth in PC12 cells. PC12 cells were cultured in the presence or absence on nonliposomal or liposomal-PureWay-C® at 5 mM for 6 hours and photographed at 320 X total magnification.

enhanced neuronal cell function (**Figure 2**) compared to nonliposomal-vitamin C. Indeed while vitamin C enhances wound healing in general [64–66], the liposomal form of vitamin C shows improved absorption into cells [16–20] and improved wound healing activity [16]. Further advances in liposomal delivery of vitamin C may be accomplished by incorporating recognition aptamer molecules into the lipid layer that can help with targeted delivery. For example, proteins that bind to epithelial cells markers may even further increase liposomal-vitamin C absorption into epithelial cells and in this manner liposomal-vitamin C delivery can be made to be more targeted using liposomes.

5. Conclusions

Vitamin C is a required nutrient for a variety of biological activities all of which require absorption from the gut and crossing lipid membrane barriers in order to arrive at the site of action. Naturally occurring vitamin C and most supplements are in the form of ascorbic acid and is hydrophilic. Vitamin C absorption in the involves active transport through the SVCT channel proteins and also through diffusion. Improved nutrient and vitamin C absorption in the gut an distribution in the body is desirable and increased efficiency in nutrition absorption and distribution improves health and overall nutrition. Vitamin C has a wide range of functions in the body including functioning as an antioxidant, a co-factor for collagen synthesis, as an immunomodulatory and as a factor for skin, cardiovascular and nervous system health including anti-cancer properties [5, 16, 23, 24]. Due to the importance of vitamin C in nearly all tissues, a great deal of research has gone into developing more bioavailable and better absorbed forms of vitamin C. Improved absorption of liposomal forms of vitamin C has been associated with improved neuronal, immunological and cardiovascular function and protection from ischemia [16–20, 49] and provides benefits in epithelial wound healing [48]. Further advances in liposomal delivery vitamin C may be accomplished by incorporating recognition molecules into the lipid layer that can help with delivery. For example, proteins that bind to epithelial cells markers may even further increase liposomal-vitamin C absorption into epithelial cells. Screening the bioactivity of these liposomal forms of vitamin C in the in vitro scratch assay is an excellent bioassay to continue to assess the efficacy of liposomal forms of vitamin C as they are developed. Caution is best taken moving forward using the liposomal approach as liposomes from drug therapies have been shown to accumulate in the macrophages of the murine bone marrow and stimulate pro-inflammatory processes and interfere with hematopoiesis [67]. This and other concerns will have to be kept in mind as the liposomal industry moves forward to provide targeted gene therapy, cancer and drug treatment and the bioactivity of liposomal LiposoMax™ nutraceuticals such as glutathione, quercetin and vitamin C [3, 4, 16].

Conflict of interest

Benjamin S. Weeks, Ph.D. has received grants and financial support from One Innovation Labs, 14,520 NW 60th Avenue, Miami Lakes, FL 33014.

Author details

Benjamin S. Weeks[1]* and Pedro P. Perez[2]

1 Department of Biology and Environmental Studies Program, Adelphi University, Garden City, NY, United States

2 One Innovation Labs, Miami Lakes, FL, United States

*Address all correspondence to: weeksbenjamins@gmail.com

References

[1] van Meer G, Voelker DR, Feigenson GW. Membrane lipids: Where they are and how they behave. Nature Reviews. Molecular Cell Biology. 2008;**9**(2):112-124. DOI: 10.1038/nrm2330

[2] Dymek M, Sikora E. Liposomes as biocompatible and smart delivery systems - the current state. Advances in Colloid and Interface Science. 2022;**309**:102757. DOI: 10.1016/j.cis.2022.102757

[3] Shah S, Dhawan V, Holm R, Nagarsenker MS, Perrie Y. Liposomes: Advancements and innovation in the manufacturing process. Advanced Drug Delivery Reviews. 2020;**154-155**:102-122. DOI: 10.1016/j.addr.2020.07.002

[4] Liu P, Chen G, Zhang J. A review of liposomes as a drug delivery system: Current status of approved products, regulatory environments, and future perspectives. Molecules. 2022;**27**(4):1372. DOI: 10.3390/molecules27041372

[5] Lykkesfeldt J, Tveden-Nyborg P. The pharmacokinetics of vitamin C. Nutrients. 2019;**11**(10):2412. Published 2019 Oct 9. DOI: 10.3390/nu11102412

[6] Washko PW, Wang Y, Levine M. Ascorbic acid recycling in human neutrophils. The Journal of Biological Chemistry. 1993;**268**(21):15531-15535

[7] Kon E, Elia U, Peer D. Principles for designing an optimal mRNA lipid nanoparticle vaccine. Current Opinion in Biotechnology. 2022;**73**:329-336. DOI: 10.1016/j.copbio.2021.09.016

[8] Dzau VJ, Mann MJ, Morishita R, Kaneda Y. Fusigenic viral liposome for gene therapy in cardiovascular diseases. Proceedings of the National Academy of Sciences of the United States of America. 1996;**93**(21):11421-11425. DOI: 10.1073/pnas.93.21.11421

[9] Son JS, Chow R, Kim H, Lieu T, Xiao M, Kim S, et al. Liposomal delivery of gene therapy for ovarian cancer: A systematic review. Reproductive Biology and Endocrinology. 2023;**21**(1):75. DOI: 10.1186/s12958-023-01125-2

[10] Ito K, Chen J, Asano T, Vaughan ED Jr, Poppas DP, Hayakawa M, et al. Liposome-mediated gene therapy in the kidney. Human Cell. 2004;**17**(1):17-28. DOI: 10.1111/j.1749-0774.2004.tb00016.x

[11] Moosavian SA, Sahebkar A. Aptamer-functionalized liposomes for targeted cancer therapy. Cancer Letters. 2019;**448**:144-154. DOI: 10.1016/j.canlet.2019.01.045. Epub 2019 Feb 11

[12] Whitt M, Buonocore L, Rose JK. Liposome-mediated transfection. Current Protocols in Immunology. 2001;**10**:10.16.1-10.16.4. DOI: 10.1002/0471142735.im1016s03

[13] Gorantla S, Wadhwa G, Jain S, Sankar S, Nuwal K, Mahmood A, et al. Recent advances in nanocarriers for nutrient delivery. Drug Delivery and Translational Research. 2022;**12**(10):2359-2384. DOI: 10.1007/s13346-021-01097-z

[14] Singh M, Devi S, Rana VS, Mishra BB, Kumar J, Ahluwalia V. Delivery of phytochemicals by liposome cargos: Recent progress, challenges and opportunities. Journal of Microencapsulation. 2019;**36**(3):215-235. DOI: 10.1080/02652048.2019.1617361

[15] Caritá AC, Fonseca-Santos B, Shultz JD, Michniak-Kohn B, Chorilli M,

Leonardi GR. Vitamin C: One compound, several uses. Advances for delivery, efficiency and stability. Nanomedicine. 2020;**24**:102117. DOI: 10.1016/j.nano.2019.102117

[16] Weeks SB et al. Vitamin C promotes wound healing: The use of in vitro scratch assays to assess Re-epithelialization. In: Physiology. London, UK: IntechOpen; 2023. DOI: 10.5772/intechopen.111517

[17] Mikirova N, Levy T, Hunninghake R. The Levels of Ascorbic Acid in Blood and Mononuclear Blood Cells after Oral Liposome-Encapsulated and Oral Non-Encapsulated Vitamin C Supplementation, Taken without and with IV Hydrocortisone. ISOM. Published January 31, 2024. Available from: https://isom.ca/article/the-levels-of-ascorbic-acid-in-blood-and-mononuclear-blood-cells-after-oral-liposome-encapsulated-and-oral-non-encapsulated-vitamin-c-supplementation-taken-without-and-with-iv-hydrocortisone/

[18] Davis JL, Paris HL, Beals JW, et al. Liposomal-encapsulated ascorbic acid: Influence on vitamin C bioavailability and capacity to protect against ischemia-reperfusion injury. Nutrition and Metabolic Insights. 2016;**9**:25-30. Published 2016 Jun 20. DOI: 10.4137/NMI.S39764

[19] Khalili A, Alipour S, Fathalipour M, et al. Liposomal and non-liposomal formulations of vitamin C: Comparison of the antihypertensive and vascular modifying activity in renovascular hypertensive rats. Iranian Journal of Medical Sciences. 2020;**45**(1):41-49. DOI: 10.30476/ijms.2019.45310

[20] Gopi S, Balakrishnan P. Evaluation and clinical comparison studies on liposomal and non-liposomal ascorbic acid (vitamin C) and their enhanced bioavailability. Journal of Liposome Research. 2021;**31**(4):356-364. DOI: 10.1080/08982104.2020.1820521

[21] Ko J, Yoo C, Xing D, et al. Pharmacokinetic analyses of liposomal and non-liposomal multivitamin/mineral formulations. Nutrients. 2023;**15**(13):3073. Published 2023 Jul 7. DOI: 10.3390/nu15133073

[22] Joseph A, Kumar D, Balakrishnan A, Shanmughan P, Maliakel B, Im K. Surface-engineered liposomal particles of calcium ascorbate with fenugreek galactomannan enhanced the oral bioavailability of ascorbic acid: A randomized, double-blinded, 3-sequence, crossover study. RSC Advances. 2021;**11**(60):38161-38171. Published 2021 Nov 26. DOI: 10.1039/d1ra06483e

[23] Padayatty SJ, Levine M. Vitamin C: The known and the unknown and goldilocks. Oral Diseases. 2016;**22**(6):463-493. DOI: 10.1111/odi.12446

[24] Lachapelle MY, Drouin G. Inactivation dates of the human and Guinea pig vitamin C genes. Genetica. 2011;**139**(2):199-207. DOI: 10.1007/s10709-010-9537-x

[25] Hemilä H. Vitamin C and infections. Nutrients. 2017;**9**(4):339. Published 2017 Mar 29. DOI: 10.3390/nu9040339

[26] Juneja D, Nasa P, Jain R. Current role of high dose vitamin C in sepsis management: A concise review. World Journal of Critical Care Medicine. 2022;**11**(6):349-363. Published 2022 Nov 9. DOI: 10.5492/wjccm.v11.i6.349

[27] Shahbaz U, Fatima N, Basharat S, et al. Role of vitamin C in preventing of COVID-19 infection, progression

and severity. AIMS Microbiology. 2022;**8**(1):108-124. Published 2022 Mar 30. DOI: 10.3934/microbiol.2022010

[28] Holford P, Carr AC, Jovic TH, et al. Vitamin C-an adjunctive therapy for respiratory infection, sepsis and COVID-19. Nutrients. 2020;**12**(12):3760. Published 2020 Dec 7. DOI: 10.3390/nu12123760

[29] Mehta N, Pokharna P, Shetty SR. Unwinding the potentials of vitamin C in COVID-19 and other diseases: An updated review. Nutrition and Health. 2022;**29**(3):415-433. DOI: 10.1177/02601060221139628

[30] Huang WY, Hong J, Ahn SI, Han BK, Kim YJ. Association of Vitamin C treatment with clinical outcomes for COVID-19 patients: A systematic review and meta-analysis. Healthcare (Basel). 2022;**10**(12):2456. Published 2022 Dec 5. DOI: 10.3390/healthcare10122456

[31] Sokary S, Ouagueni A, Ganji V. Intravenous ascorbic acid and lung function in severely ill COVID-19 patients. Metabolites. 2022;**12**(9):865. Published 2022 Sep 14. DOI: 10.3390/metabo12090865

[32] Webb AL, Villamor E. Update: Effects of antioxidant and non-antioxidant vitamin supplementation on immune function. Nutrition Reviews. 2007;**65**(5):181-217. DOI: 10.1111/j.1753-4887.2007.tb00298.x

[33] Manning J, Mitchell B, Appadurai DA, et al. Vitamin C promotes maturation of T-cells. Antioxidants and Redox Signaling. 2013;**19**(17):2054-2067. DOI: 10.1089/ars.2012.4988

[34] Carr AC, Maggini S. Vitamin C and immune function. Nutrients. 2017;**9**(11):1211. Published 2017 Nov 3. DOI: 10.3390/nu9111211

[35] Molina N, Morandi AC, Bolin AP, Otton R. Comparative effect of fucoxanthin and vitamin C on oxidative and functional parameters of human lymphocytes. International Immunopharmacology. 2014;**22**(1):41-50. DOI: 10.1016/j.intimp.2014.06.026

[36] Ströhle A, Hahn A. Vitamin C und Immunfunktion [Vitamin C and immune function]. Medizinische Monatsschrift für Pharmazeuten. 2009;**32**(2):49-56

[37] Moretti M, Rodrigues ALS. Functional role of ascorbic acid in the central nervous system: A focus on neurogenic and synaptogenic processes. Nutritional Neuroscience. 2022;**25**(11):2431-2441. DOI: 10.1080/1028415X.2021.1956848

[38] Pullar JM, Carr AC, Vissers MCM. The roles of vitamin C in skin health. Nutrients. 2017;**9**(8):866. Published 2017 Aug 12. DOI: 10.3390/nu9080866

[39] Bechara N, Flood VM, Gunton JE. A systematic review on the role of vitamin C in tissue healing. Antioxidants (Basel). 2022;**11**(8):1605. Published 2022 Aug 19. DOI: 10.3390/antiox11081605

[40] Parsons KK, Maeda N, Yamauchi M, Banes AJ, Koller BH. Ascorbic acid-independent synthesis of collagen in mice. American Journal of Physiology. Endocrinology and Metabolism. 2006;**290**(6):E1131-E1139. DOI: 10.1152/ajpendo.00339.2005

[41] Tajima S, Pinnell SR. Ascorbic acid preferentially enhances type I and III collagen gene transcription in human skin fibroblasts. Journal of Dermatological Science. 1996;**11**(3):250-253. DOI: 10.1016/0923-1811(95)00640-0

[42] Duarte TL, Cooke MS, Jones GD. Gene expression profiling reveals new protective roles for vitamin C in human skin cells. Free

Radical Biology and Medicine. 2009;**46**(1):78-87. DOI: 10.1016/j.freeradbiomed.2008.09.028

[43] Lund CC, Crandon JH. Ascorbic acid and human wound healing. Annals of Surgery. 1941;**114**(4):776-790. DOI: 10.1097/00000658-194110000-00019

[44] Yun IS, Yoo HS, Kim YO, Rah DK. Improved scar appearance with combined use of silicone gel and vitamin C for Asian patients: A comparative case series. Aesthetic Plastic Surgery. 2013;**37**(6):1176-1181. DOI: 10.1007/s00266-013-0210-5

[45] Silverstein RJ, Landsman AS. The effects of a moderate and high dose of vitamin C on wound healing in a controlled Guinea pig model. The Journal of Foot and Ankle Surgery. 1999;**38**(5):333-338. DOI: 10.1016/s1067-2516(99)80004-0

[46] Boo YC. Ascorbic acid (vitamin C) as a cosmeceutical to increase dermal collagen for skin antiaging purposes: Emerging combination therapies. Antioxidants (Basel). 2022;**11**(9):1663. Published 2022 Aug 26. DOI: 10.3390/antiox11091663

[47] Mitmesser SH, Ye Q, Evans M, Combs M. Determination of plasma and leukocyte vitamin C concentrations in a randomized, double-blind, placebo-controlled trial with Ester-C(®). Springerplus. 2016;**5**(1):1161. Published 2016 Jul 25. DOI: 10.1186/s40064-016-2605-7

[48] Weeks BS, Perez PP. A novel vitamin C preparation enhances neurite formation and fibroblast adhesion and reduces xenobiotic-induced T-cell hyperactivation. Medical Science Monitor. 2007;**13**(3):BR51-BR58

[49] Weeks BS, Perez PP. Absorption rates and free radical scavenging values of vitamin C-lipid metabolites in human lymphoblastic cells. Medical Science Monitor. 2007;**13**(10):BR205-BR210

[50] Pancorbo D, Vazquez C, Fletcher MA. Vitamin C-lipid metabolites: Uptake and retention and effect on plasma C-reactive protein and oxidized LDL levels in healthy volunteers. Medical Science Monitor. 2008;**14**(11):CR547-CR551

[51] Anderson KE, Eliot LA, Stevenson BR, Rogers JA. Formulation and evaluation of a folic acid receptor-targeted oral vancomycin liposomal dosage form. Pharmaceutical Research. 2001;**18**(3):316-322. DOI: 10.1023/a:1011002913601

[52] Ling SS, Yuen KH, Magosso E, Barker SA. Oral bioavailability enhancement of a hydrophilic drug delivered via folic acid-coupled liposomes in rats. The Journal of Pharmacy and Pharmacology. 2009;**61**(4):445-449. DOI: 10.1211/jpp/61.04.0005

[53] Jalali-Jivan M, Rostamabadi H, Assadpour E, et al. Recent progresses in the delivery of β-carotene: From nano/microencapsulation to bioaccessibility. Advances in Colloid and Interface Science. 2022;**307**:102750. DOI: 10.1016/j.cis.2022.102750

[54] Rovoli M, Pappas I, Lalas S, Gortzi O, Kontopidis G. In vitro and in vivo assessment of vitamin a encapsulation in a liposome-protein delivery system. Journal of Liposome Research. 2019;**29**(2):142-152. DOI: 10.1080/08982104.2018.1502314

[55] Dałek P, Drabik D, Wołczańska H, et al. Bioavailability by design - vitamin D_3 liposomal delivery vehicles. Nanomedicine. 2022;**43**:102552. DOI: 10.1016/j.nano.2022.102552

[56] Nowak JK, Sobkowiak P, Drzymała-Czyż S, et al. Fat-soluble vitamin supplementation using liposomes, cyclodextrins, or medium-chain triglycerides in cystic fibrosis: A randomized controlled trial. Nutrients. 2021;**13**(12):4554. Published 2021 Dec 20. DOI: 10.3390/nu13124554

[57] Cheng CY, Barro L, Tsai ST, et al. Epigallocatechin-3-Gallate-loaded liposomes favor anti-inflammation of microglia cells and promote neuroprotection. International Journal of Molecular Sciences. 2021;**22**(6):3037. Published 2021 Mar 16. DOI: 10.3390/ijms22063037

[58] Kala SG, Chinni S. Bioavailability enhancement of vitamin E TPGS liposomes of nintedanib esylate: Formulation optimization, cytotoxicity and pharmacokinetic studies. Drug Delivery and Translational Research. 2022;**12**(11):2856-2864. DOI: 10.1007/s13346-022-01148-z

[59] Ahmad A, Vaghasiya K, Kumar A, et al. Enema based therapy using liposomal formulation of low molecular weight heparin for treatment of active ulcerative colitis: New adjunct therapeutic opportunity. Materials Science and Engineering. C, Materials for Biological Applications. 2021;**121**:111851. DOI: 10.1016/j.msec.2020.111851

[60] Tinsley GM, Harty PS, Stratton MT, Siedler MR, Rodriguez C. Liposomal mineral absorption: A randomized crossover trial. Nutrients. 2022;**14**(16):3321. Published 2022 Aug 13. DOI: 10.3390/nu14163321

[61] Shariare MH, Pinky NJK, Abedin J, Kazi M, Aldughaim MS, Uddin MN. Liposomal drug delivery of *Blumea lacera* leaf extract: In-vivo Hepatoprotective effects. Nanomaterials (Basel). 2022;**12**(13):2262. Published 2022 Jun 30. DOI: 10.3390/nano12132262

[62] Zhang J, Li X, Huang L. Anticancer activities of phytoconstituents and their liposomal targeting strategies against tumor cells and the microenvironment. Advanced Drug Delivery Reviews. 2020;**154-155**:245-273. DOI: 10.1016/j.addr.2020.05.006

[63] Dutta S, Moses JA, Anandharamakrishnan C. Encapsulation of nutraceutical ingredients in liposomes and their potential for cancer treatment. Nutrition and Cancer. 2018;**70**(8):1184-1198. DOI: 10.1080/01635581.2018.1557212

[64] Chaitrakoonthong T, Ampornaramveth R, Kamolratanakul P. Rinsing with L-ascorbic acid exhibits concentration-dependent effects on human gingival fibroblast In vitro wound healing behavior. International Journal of Dentistry. 2020;**2020**:4706418. Published 2020 Mar 21. DOI: 10.1155/2020/4706418

[65] Yi Y, Wu M, Zhou X, et al. Ascorbic acid 2-glucoside preconditioning enhances the ability of bone marrow mesenchymal stem cells in promoting wound healing. Stem Cell Research and Therapy. 2022;**13**:1, 119. Published 2022 Mar 21. DOI: 10.1186/s13287-022-02797-0

[66] Chong WT, Tan CP, Cheah YK, Lai OM. In-vitro and in-vivo evaluations of tocotrienol-rich nanoemulsified system on skin wound healing. PLoS One. 2022;**17**(5). DOI: 10.1371/journal.pone.0267381

[67] Li Y, Yao R, Ren M, Yuan K, Du Y, He Y, et al. Liposomes trigger bone marrow niche macrophage "foam" cell formation and affect hematopoiesis in mice. Journal of Lipid Research. 2022;**63**(10):100273. DOI: 10.1016/j.jlr.2022.100273